PHOSPHORIC ACID INDUSTRY - PROBLEMS AND SOLUTIONS

Edited by **Michael Schorr Wiener**
and **Benjamin Valdez**

Phosphoric Acid Industry - Problems and Solutions
http://dx.doi.org/10.5772/63661
Edited by Michael Schorr Wiener and Benjamin Valdez

Contributors

José Domingos Fontana, Marcela Tiboni, Heidegrid Siebert Koop, Mohamed Jemal, Richard Thilagaraj, P. Venu Babu, Uma Selvaraj, Gouri Chaudhuri, Khatereh Bahrpaima, Ernesto Brunet, Hussein Alhendawi, Elena Rodríguez-Payán, Juan Pedro Bolivar, Rafael Gacia Tenorio, Fernando Mosqueda, Juan Mantero, José Luis Guerrero-Márquez, Guillermo Manjon, Michael Schorr

Published by InTech

Janeza Trdine 9, 51000 Rijeka, Croatia

Publishing Process Manager Iva Simcic
Technical Editor SPi Global
Cover InTech Design team

First published July, 2017
Printed in Croatia
Legal deposit, Croatia: National and University Library in Zagreb

Additional hard copies can be obtained from orders@intechopen.com

Phosphoric Acid Industry - Problems and Solutions, Edited by Michael Schorr Wiener and Benjamin Valdez
p. cm.
Print ISBN 978-953-51-3353-7
Online ISBN 978-953-51-3354-4

For EU product safety concerns:
IN TECH d.o.o., Prolaz Marije Krucifikse Kozulić 3, 51000 Rijeka, Croatia, info@intechopen.com or visit our website at intechopen.com.

Meet the editors

Michael Schorr Wiener is a professor and doctor honoris causa at the Institute of Engineering, Universidad Autonoma de Baja California, Mexicali, Baja California, México. He has a BSc degree in chemistry and an MSc degree in materials engineering from the Technion-Israel Institute of Technology. He has published 406 scientific and technical articles on materials and corrosion in English, Spanish, and Hebrew, including books by InTech. He has worked as a corrosion consultant and professor in Israel, the United States, Latin America, Spain, South Africa, and France. He has labored for the phosphoric acid industry in Israel, Europe, Venezuela, and South Africa during 1960–1998 and in Mexico and Brazil during 1998–2015. He is a member of the National System of Researchers in Mexico.

Benjamin Valdez was the director of the Institute of Engineering during 2006–2013, Universidad Autonoma de Baja California, Mexicali, Baja California, México. He has a BSc degree in chemical engineering, has an MSc degree and a PhD degree in chemistry, and is a member of the Mexican Academy of Science and the National System of Researchers in Mexico. He was the guest editor of Corrosion Reviews, in which he produced two special issues on corrosion control in geothermal plants and the electronic industry. He is a full professor at the University of Baja California. His activities include corrosion research, consultancy, and control in industrial plants and environments. In August 2013, Mexico, he received a "Distinguished Service Award" from NACE International and Mexico Section.

Contents

Preface

Phosphoric acid is an important industrial acid that is utilized for manufacturing phosphatic fertilizers and industrial products, for pickling and posterior treatment of steel surfaces to prevent corrosion, for ensuring appropriate paint adhesion, and for the food and beverages industry, e.g., cola-type drinks to impart taste and slight acidity and to avoid iron sedimentation. A singular application is the manufacture of artificial apatite for coating on stainless steel orthopedic implants in the human body.

Two main processes are used to produce this acid: the wet process PA (WPA) by treatment with H_2SO_4 and a relatively novel process by treatment with HCl followed by separation of PA applying solvent extraction technology. The composition and purity of the PA obtained by these processes depend upon the mineral and chemical composition of the phosphate rock and the acid used.

The global, economic, and social importance of the phosacid industry is evident by the realization of the International Symposium on Innovation and Technology in the Phosphate Industry, with outstanding success (SYMPHOS, 2013, 2015).

This industry is spread out in countries of four continents—Asia, Africa, America, and Europe—which operate mines and production plants and produce fertilizers. Phosacid is one of the most widely known acids. It is used in metallurgy, medicine, dentistry, and cosmetics; as an electrolyte for fuel cells; and as a quality promoter in the food industry.

This volume is well organized displaying the main problems affecting the industry and the solutions proposed in different fields.

In Chapter 1, M. Jamal et al. deal with the thermochemistry and kinetic reaction of apatite with acid solutions. W. Richard et al. explain how alkaline phosphatase is applied for bioremediation of heavy metals from industrial and nuclear wastes in Chapter 2. The theoretical aspects of polysaccharide depolymerization are assessed by J Domingos et al.

M. Schorr and B. Valdez review the world phosphate rock market and the producers in several countries. The environmental problems generated by the phosphoric acid industry in Huelva, Spain, are described by Bolivar et al in Chapter 5. The technological application of phosphate-based materials is presented in Chapter 6. The purification of phosacid is shown in Chapter 7.

This multiauthored book offers a wide spectrum of basic knowledge and practical experience affecting the wide phosphoric acid industry. These learned specialists, producers, and researchers manifest different approaches and orientation that will be useful for the whole industry. The global phosacid market and its many phosphate derivatives are expanding worldwide; this trend is expected to continue in the next years, thus producing innovative products.

Dr. Michael Schorr Wiener and Dr. Benjamin Valdez Salas
Institute of Engineering
University of Baja California, Mexico

Section 1

Theoretical Aspects

Chapter 1

Thermochemistry and Kinetics of the Reactions of Apatite Phosphates with Acid Solutions (II)

Mohamed Jemal

Additional information is available at the end of the chapter

http://dx.doi.org/10.5772/68087

Abstract

The principal material in the phosphate ores is composed of calcium fluorapatite $Ca_{10}(PO_4)_6F_2$, in which the various components have been partially substituted by magnesium, sodium, carbonate, and hydroxyl ions. These substitutions affect the stability of the material and its reactivity toward the acid attack. The present chapter reports the influence of carbonates and magnesium on these properties. Using different calorimeters, dissolution experiments of carbonated and noncarbonated Ca and Ca/Mg apatites were carried out in acid solutions leading to thermochemical quantities. The results show that substitution of carbonate for F ions in the channel (to get A-type carbonate F-apatites) results in increasing the stability of the edifice, while substitution of CO_3 for PO_4 in fluor- or hydroxyapatites (to get B-type apatites) leads to a decrease in stability. The latter phenomenon was also observed when substituting magnesium for calcium in F-apatites. The presence of the former in the apatite structure results in an increase of the speed of dissolution in acid solution that is enhanced when carbonate is also replacing phosphate groups. Dissolution mechanism of synthesized Ca/Mg F-apatites seems to be a one-step process, while dissolution of a Gafsa (TN) natural phosphate to get superphosphate fertilizer is more complex.

Keywords: carbonate apatites, enthalpy of formation, Gibbs free energy, acid attack, phosphate ore, kinetic models

1. Introduction

The "wet" manufacturing process of phosphoric acid is based on the attack of natural phosphates by nitric, hydrochloric, or sulfuric acid or the mixture of the latter with diluted phosphoric acid solution. Super phosphate fertilizers are produced industrially through the reaction of phosphoric acid with natural phosphates.

According to Pierre Becker [1], "the most important (mineral) for the phosphoric acid industry is the apatite group: $(Ca,Na_2,Mg)_{10}(PO_4)_{6-x}(CO_3)_xF_y(F,OH)_{2-y}$." This formula is representative of a large variety of compounds in which the compositions in the major constituents, that is, calcium, phosphorus, and fluorine, lie in the ranges 29.3–53.6, 23.8–40.3, and 1.3–4.1 w% CaO, P_2O_5 and F, respectively (P. Becker). The deposit compounds are probably the most stable ones appearing in their respective geological conditions of temperature, pressure, and composition in different elements.

It would be interesting to apprehend the contribution of each component in the stability and reactivity of the natural products. However, as it is impossible to reproduce the formation conditions of the latter, one can proceed by a progressive approach which consists in determining these properties for synthetic limit products then for their "derivative" which can be obtained by introduction of various elements or entities in the former. Thermodynamics allows to quantify the stability of the compounds through the determination of their Gibbs free energy of formation and gives a comparison between compounds having molar weights in the same order of magnitude. Kinetic studies of the acid attack allow to characterize the reactivity of the compounds in terms of rapidity, to propose an attack mechanism and highlight the influence of some parameters on the reaction rate.

Because of the predominance in the ore of the components cited above, Ca-fluorapatite (Ca-FAP), $Ca_{10}(PO_4)_6F_2$, has been considered as the more representative compound of natural phosphates and has been widely studied under various aspects. This is also because of the similarity of their crystal structures. Ca-hydroxyapatite (Ca-HAP), $Ca_{10}(PO_4)_6(OH)_2$, comes in the second position and constitutes the model compound in studying the properties of the mineral component of calcified tissues (enamel, dentine, and bones).

Ca-FAP seems to be crystallographically characterized for the first time by Leonhardt in 1923 (cited in ref. [2]). The structure has been confirmed in 1930 [2, 3]. Since then a wide varieties of works have been performed [4], but the first chapter related to thermochemistry of this compound has been published by Gottschall in 1958 [5] who dissolved Ca-FAP in nitric acid and determined its formation enthalpy. At the end of the 1990s, a lot of thermochemistry works started on Ca-FAP and analogous compounds containing Ba, Sr, Mg, Cd, and Pb, in order to determine their dissolution energies and to deduce their formation enthalpies. In absence of formation entropies, these results allowed to compare the stability of the compounds based on the formation enthalpies and not on the Gibbs free energies, as required. Then the entropy factor has been estimated by a calculation procedure combining literature data [6]. This allowed to propose a more realistic scale of stability of these compounds. The same calculation procedure has been applied for hydroxy and chlor-apatites containing the metal reported above. All these results are gathered in reference [7]. Substitutions of calcium by alkali earth metals, Cd or Pb have also been considered thermochemically and have been widely published in literature. In the phosphate ore, these substitutions are minor in comparison to carbonate substitutions. This chapter deals with the stability and reactivity of some carbonate apatites and reports thermodynamic and kinetic studies of the attack of a sample ore from Gafsa (Tunisia) by phosphoric solution having different composition.

http://dx.doi.org/10.5772/68087

2. Carbonate substitutions in the apatites

The model compound, Ca-FAP, crystallizes in the hexagonal system with space group $P6_3/m$. The structure has been refined from diffractometer single-crystal X-ray data in 1972 [4]. The corners of the hexagonal unit cell are occupied by calcium ions, called Ca(1), which are also present at the center of the *c*-axis parameter. These columns of Ca^{2+} ions are linked together by PO_4 tetrahedron groups that are positioned along both sides of the hexagonal faces. The other six calcium, called Ca(2), form two triangles rotated by 60° from each other around the *c*-axis and are located at ¼ and ¾ positions along this axis, forming a channel. In FAP, the center of each triangle is occupied by a F ion, while in hydroxy and chlorapatites, hydroxyl, and chloride ions are above this position. **Figure 1** shows the projection of the FAP lattice on the a, b plane.

Carbonate can substitute for F (or OH) along the channel or for PO_4 groups, leading to "A" or "B" type carbonate apatites, respectively. A-type carbonate apatites are synthesized by solid gas/reaction at high temperature while the B-type ones are precipitated in aqueous solution. AB-type carbonate apatites have either been prepared by solid/gas reaction performed on rich carbonate B-type ones [4], or by precipitation in conditions similar to that used for the synthesis of Ca-deficient hydroxyapatites [8]. Each type of compounds are characterized crystallographically by specific displacements of some rays on the X-ray profiles and spectroscopically by the appearance of special bands on the IR recordings. They also differ in some properties among them, the thermodynamic ones, as seen below.

Fluoride (or hydroxyl) replacement mechanism leading to A-type carbonate apatites is commonly supposed to be accompanied by the appearance of an anion vacancy along the channel. This seems to be accepted without any ambiguity while in the B-type ones, substitution of PO_4 by CO_3 has been subjected to a large number of suggestions resulting from various methods of treatment of the analytical results.

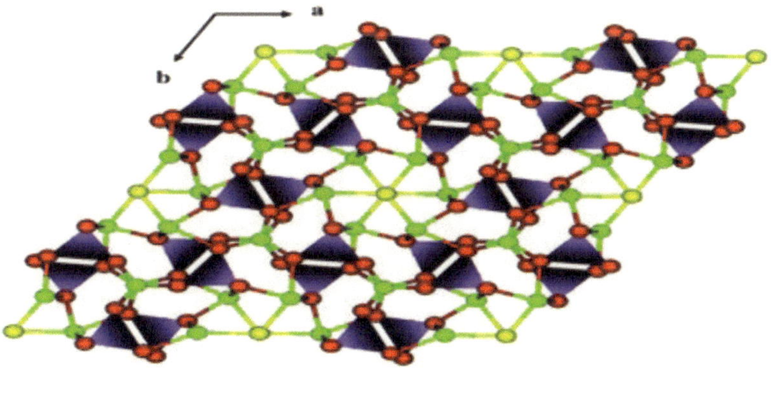

Figure 1. Projection on a,b plane of the Ca-FAP structure showing Ca(1) in green, O in red, PO_4 in blue, and F in yellow at the center of Ca(1) hexagons. Ca(2) are not represented .

In 1993, De Mayer and Verbeeck [9] gathered the various substitution mechanisms that can be considered in B-type carbonate hydroxyapatites containing sodium. These mechanisms are as follows:

$$\begin{aligned}
&(I)\quad Ca^{2+} + PO_4^{3-} \leftrightarrow V^{Ca} + CO_3^{2-} + V^{OH} \\
&(II)\quad Ca^{2+} + 2PO_4^{3-} \leftrightarrow V^{Ca} + 2CO_3^{2-} \\
&(III)\quad Ca^{2+} + PO_4^{3-} \leftrightarrow Na^{+} + V^{OH} \\
&(IV)\quad Ca^{2+} + OH^{-} \leftrightarrow Na^{+} + V^{OH} \\
&(V)\quad PO_4^{3-} \leftrightarrow CO_3^{2-} + OH^{-},
\end{aligned}$$

where V^{Ca} and V^{OH} stand for vacancies of Ca^{2+} and OH^{-}

Replacement mechanisms I– IV are accompanied by creation of a vacancy in the Ca or OH site, while in mechanism V, PO_4 is replaced by the tandem (CO_3OH). The latter mechanism has been suggested considering the analogy with the fluoride analogous apatites in which the chemical analysis reveals the presence of more fluoride than in free carbonate fluorapatite.

Suppose *a, b, c, d,* and *e* are the respective contributions of these mechanisms, the general formula of the apatite will be:

$$Ca_{10-(a+b+c+d)}(Na)_{(c+d)}(PO_4)_{6-(a+2b+c+e)}(CO_3)_{(a+2b+c+e)}(OH)_{2-(a+c+d-e)} \tag{1a}$$

In Na free B-type hydroxyapatites, only mechanisms I, II, and V have to be considered. So "*c*" and "*d*" factors should be nil and the general formula is derived as:

$$Ca_{10-(a+b)}(PO_4)_{6-(a+2b+e)}(CO_3)_{(a+2b+e)}(OH)_{2-(a-e)} \tag{1b}$$

For the fluoride homologous compounds the formula becomes as:

$$Ca_{10-(a+b)}(PO_4)_{6-(a+2b+e)}(CO_3)_{(a+2b+e)}F_{2-(a-e)} \tag{1c}$$

This formula is equivalent to that proposed in 1963 by Kühl and Nebergall who "appear to have been the first to propose the formula: $Ca_{10-x+y}(PO_4)_{6-x}(CO_3)_xF_{2-x+2y}$" [10] with $x = a + 2b + e$ and $y = b + e$. Formula (1c) supposes that among the "$a + 2b + e$" carbonates introduced in fluorapatite "*e*" are associated to F as (CO_3F) entities and the remaining "$a + 2b$" are accompanied by the creation of "$a + b$" vacancies in the Ca sites.

3. Thermochemistry of some A-type carbonate apatites

Thermochemistry dissolution of A- and B-type carbonate apatites has been undertaken in the beginning considering experiments performed with low values of the solid/liquid ratio. This led to a simple dissolution without the appearance of any precipitate. As reported, the latter appears with higher values of this ratio.

http://dx.doi.org/10.5772/68087

3.1. Enthalpies of dissolution in nitric acid solution

Calcium, strontium, and barium A-type carbonate hydroxyapatites have been prepared by solid/gas carbonatation of the corresponding hydroxyapatites, according to the general reaction scheme:

$$M_{10}(PO_4)_6(OH)_2 + xCO_2(g) \rightarrow M_{10}(PO_4)_6(OH)_{2-2x}(CO_3)_x + xH_2O(g)$$

where M stands for calcium, strontium, or barium.

The starting solids were heated in pure CO_2 flowing gas at temperature between 700 and 1000°C during half an hour to 6 days. By varying the temperature and time they were progressively carbonated until full substitution of hydroxyl ions leading to pure A-carbonate apatites. The synthesized products were characterized by X-ray and IR analyses. All of them showed the IR bands characterizing such kinds of apatites and exhibit a nonmonotonous slight variation of "*c*" parameter, while "*a*" parameter and lattice volume increase appreciably over the carbonate content. The carbonate amount was determined by C-H-N analysis (for Ca-bearing compounds [11]), coulometry, and Rietveld processing (for Sr compounds [12]) or thermogravimetrically (for Ba compounds [13])

The solids were then dissolved in a 3 or 9 wt% nitric acid solution using a home-made isoperibol calorimeter. The device was previously described in detail [14]. Carbon dioxide gas was continuously bubbled in the liquid phase to avoid its retention after dissolution of carbonated products. The heat effect produced in the calorimetric cell was manifested by the variation of the temperature of the reaction medium which was detected by a thermistance probe acting as one of the four arms of a Wheatstone bridge. The balance voltage was amplified by a Keithley multimeter and then recorded over time. The solid to be dissolved or the liquid to be diluted was carefully introduced in a thin-walled glass bulb that was manufactured at one of the extremities of a Pyrex tube (5 mm diameter). Experiment started by searching a quasi-steady state in which the electrical current was practically null and the baseline deviated slightly from the vertical line. The reaction was started by breaking the bulb using a thin glass baton and the heat effect dissipated in the medium results in a deviation of the signal, which then became parallel to the previous baseline. The shift between the baselines is proportional to the corresponding heat effect. More details on the calibration and heat effect determination are exposed in Ref. [12]. Experimental results processing and errors calculation are developed in Ref. [13].

Figures 2–4 show the variation of the molar energies of solution for Ca, Sr, and Ba compounds as functions of the amount of carbonate in the lattice.

Depending on the cation, the enthalpy of solution decreases (Sr in 9 wt% HNO_3), increases (Ba in 3% HNO_3), or shows a bell-shaped curve (Ca in 9 wt% HNO_3).

3.2. Formation enthalpies of A-type carbonate apatites

The solution enthalpy of a particular compound can be involved in a hypothetical reaction containing other compounds for which the enthalpies of formation are previously tabulated.

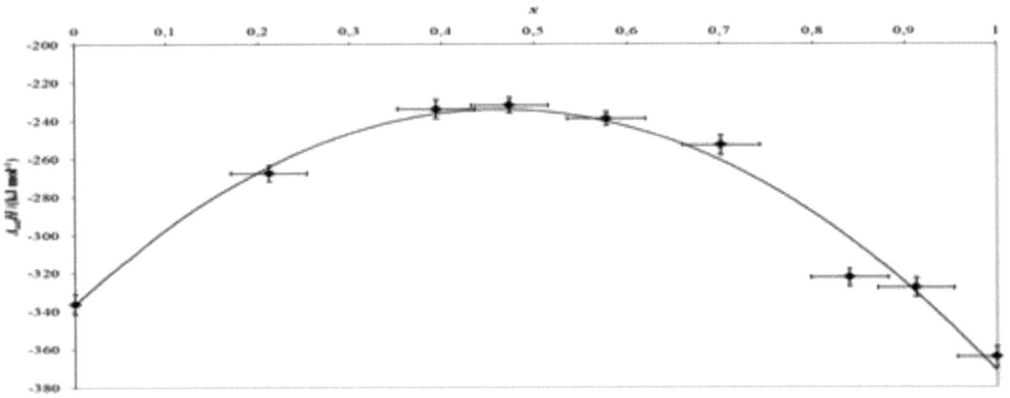

Figure 2. Standard molar enthalpy of solution at 298 K in 9 wt% nitric acid solution for $Ca_{10}(PO_4)_6(CO_3)_x(OH)_{2-2x}$ over x [11].

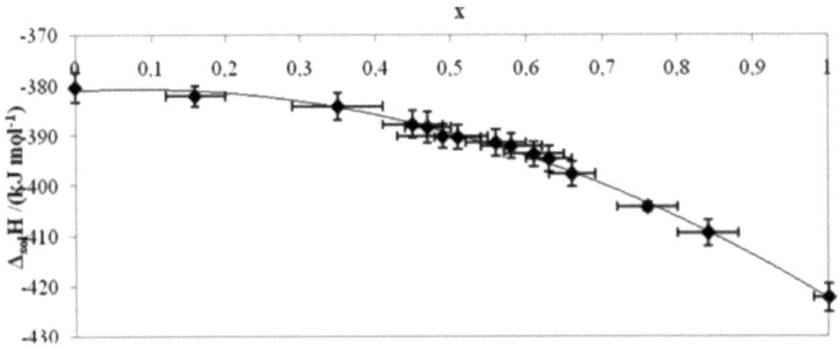

Figure 3. Standard molar enthalpy of solution at 298 K in 9 wt% nitric acid solution for $Sr_{10}(PO_4)_6(CO_3)_x(OH)_{2-2x}$ over x [12].

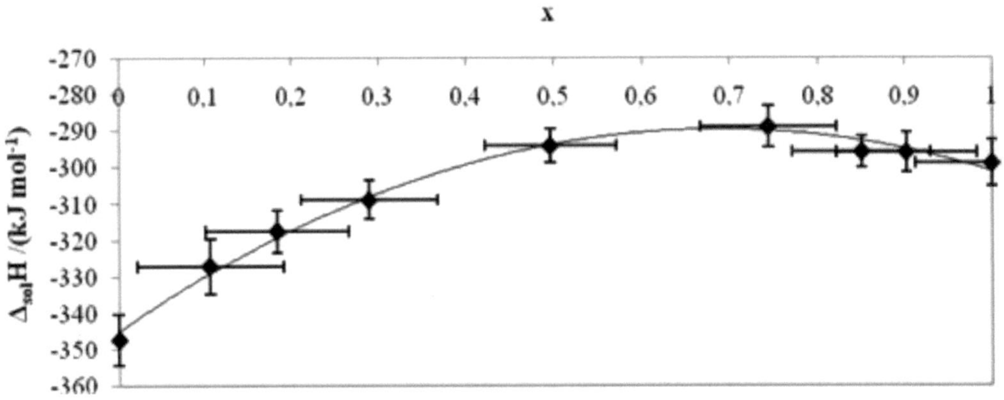

Figure 4. Standard molar enthalpy of solution at 298 K in 3 wt% nitric acid solution for $Ba_{10}(PO_4)_6(CO_3)_x(OH)_{2-2x}$ over x [13].

The difference between the enthalpies of solution of the compounds in both sides of that reaction leads to the formation enthalpy of the product in concern. Example, $M_{10}(PO_4)_6(CO_3)_x(OH)_{2\text{-}2x}$ can be involved in a reaction containing $MCO_{3(sd)}$, $M(OH)_{2\ (sd)}$, and $M_3(PO_4)_{2(sd)}$.

The dissolution reactions or their reverse are schematized as follows. Their "sum" leads to the final reaction.

$$10M(NO_3)_2 + 6H_3PO_4 + xCO_{2(g)} + (2 - x)H_2O \rightarrow M_{10}(PO_4)_6(OH)_{(2-2x)}(CO_3)_{x(sd)} + 20HNO_3 \quad (1)$$

$$x[MCO_{3(sd)} + 2HNO_3 \rightarrow M(NO_3)_2 + CO_{2(g)} + H_2O] \quad (2)$$

$$(1 - x)[M(OH)_{2(sd)} + 2HNO_3 \rightarrow M(NO_3)_2 + 2H_2O] \quad (3)$$

$$3M_3(PO_4)_{2(sd)} + 18HNO_3 \rightarrow 9M(NO_3)_2 + 6H_3PO_4 \quad (4)$$

$$xMCO_{3(sd)} + (1 - x)M(OH)_{2(sd)} + 3M_3(PO_4)_{2(sd)} \rightarrow M_{10}(PO_4)_6(OH)_{(2-2x)}(CO_3)_{x(sd)}$$

In addition to the dissolution reaction of the apatite (Step 1), this scheme includes the dissolution of x moles of M-carbonate (Step 2), $(1 - x)$ moles of M-hydroxide (Step 3), and 3 moles of trimetallic phosphate (Step 4). Their corresponding enthalpies were measured in the same solvent under similar conditions as for the apatites. The results are reported in Ref. [11] for calcium, [12] for strontium, and [13] for barium, together with the formation of enthalpies of the corresponding reactants. The latter were picked from the literature. **Figure 5** gathers the standard formation enthalpies over x for the three alkali earth metals. One can notice a monotonic variation for Sr and Ba compounds and a minimum at about x = 0.6–0.7 for Ca ones. Sr carbonate apatites seem to be less stable than the others.

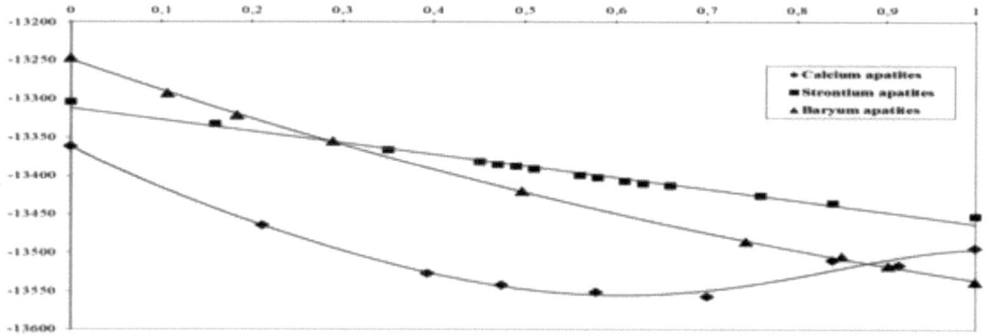

Figure 5. Comparative results of the standard formation enthalpy at temperature T = 298.15 K and pressure p = 0.1 MPa of carbonate hydroxyapatites $M_{10}(PO_4)_6(OH)_{(2\text{-}2x)}(CO_3)_x$, where M = Ca, Sr, and Ba [11].

3.3. Enthalpy of mixing in the solid state

$M_{10}(PO_4)_6(OH)_{(2-2x)}(CO_3)_x$ can be considered as a solid solution obtained by mixing $M_{10}(PO_4)_6CO_3$ $_{(sd)}$ and $M_{10}(PO_4)_6(OH)_{2(sd)}$ according to the following scheme:

$$x\ M_{10}(PO_4)_6(CO_3)_1 + (1-x)\ M_{10}(PO_4)_6(OH)_2 \rightarrow M_{10}(PO_4)_6(OH)_{(2-2x)}(CO_3)_x$$

The molar enthalpy of mixing can be determined from the solution enthalpies of the reactants and product as follows:

$$\Delta_{\text{mix}}H^\circ(x)/(kJmol^{-1}) = -\Delta_{\text{sol}}H^\circ(x) + x\Delta_{\text{sol}}H^\circ(x=1) + (1-x)\Delta_{\text{sol}}H^\circ(x=0)$$

This quantity can also be deduced from **Figure 5** by considering the difference between the enthalpy of formation of the solid solution for a given composition and that deduced from the straight line joining the points corresponding to the limit products. The later quantity equals the enthalpy of formation of the heterogeneous mixture of the limit products. **Figure 6** shows the results for calcium and barium compounds. For strontium, **Figure 5** shows a linear variation over "x" and so the mixing enthalpy is null in the all composition range, indicating that for that element, the solid solution of the limit products is an athermal one. This means that the energy bonds between the entities in the carbonate product are of the same order of magnitude as their corresponding ones in the hydroxyl compound.

3.4. Estimation of the Gibbs free energy of the A-carbonate apatites

The stability of compounds having similar chemical formulae can be estimated considering in a first approximation numerical values of their formation enthalpies. The lower the value, the more stable is the corresponding compound. This way of doing supposes the formation entropy as zero. This quantity is negative and this assumption supposes zero as the upper limit of $\Delta_r S^\circ$. However, it is possible to seek a negative upper limit and set up a more realistic stability scale, taking into account the literature data. This is done as follows.

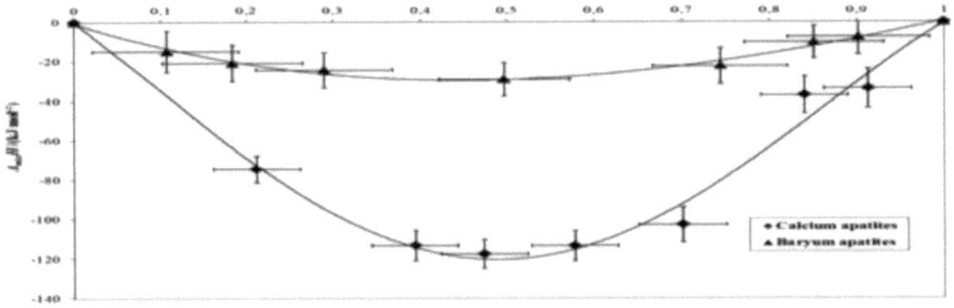

Figure 6. Variation of the mixing enthalpy of the solid solution *versus* "x" at temperature $T = 298.15$ K and pressure $p = 0.1$ MPa for calcium and barium type A CO_3-apatites.

http://dx.doi.org/10.5772/68087

Carbonate apatite having the general formula $M_{10}(PO_4)_6(OH)_{(2\text{-}2x)}(CO_3)_x$ can be obtained according the following reaction:

$$3/2\ P_4O_{10} + 10\ MO + xCO_2(g) + (1 - x)H_2O(l) \rightarrow M_{10}(PO_4)_6(OH)_{(2-2x)}(CO_3)_x$$

The standard entropy of this reaction at T° (298K), $\Delta_r S°$, can be expressed as a function of the standard entropies at T° of the reactants and product as:

$$\Delta_r S° = S°(CO3Ap)–3/2\ S°(P_4O_{10})–10.S°(MO)–x.S°(CO_2)–(1 - x).S°(H_2O)$$

Because of the increase in disorder accompanying this reaction, its entropy should be negative and so

$$S°(CO3Ap) < 3/2\ S°(P_4O_{10}) + 10.S°(MO) + x.S°(CO_2) + (1 - x).S°(H_2O).$$

The absolute entropy of the CO3-apatite can be derived from the entropy of its formation, $\Delta_r S°$ (CO3Ap), and the absolute entropies of the elements (M, P, C, O_2, and H_2) in their most stable states as:

$$S°(CO3Ap) = \Delta_f S°(CO3Ap) + 10.S°(M) + 6.S°(P) + x.S°(C) + (13 - x/2).S°(O_2) + (1 - x).S°(H_2)$$

Taking into account the inequality above, one can derive the following relationship:

$$\begin{aligned}\Delta_r S°(CO3Ap) < {} & 3/2\ S°(P_4O_{10})–6.S°(P) + 10.[S°(MO) - S°(M)] + x.[S°(CO_2) - S°(C)] \\ & + (1 - x).[S°(H_2O) - S°(H_2)]–(13 - x/2).S°(O_2)\end{aligned}$$

The second term of this inequality is negative and constitutes the upper limit value of the formation entropy, instead of zero. If one supposes $\Delta_r S°$(CO3Ap) as equal to this limit, one can calculate a more realistic value of the Gibbs free energy of formation.

Taking into account the literature data for P_4O_{10} [15] and for the other elements and compounds [16], this assumption led to the general formula for the formation entropy as:

$$\Delta_r S°(CO3Ap) = -2992.23 + 165.99.x + 10.[S°(MO)–S°(M)]$$

And so the standard formation entropy of the apatite is linear over the carbonate content (x) with an intercept value depending on M (= Ca, Sr, or Ba).

These results have been associated with the formation enthalpies previously published to lead to the Gibbs free energies of formation of the Ca, Sr, and Ba A-carbonate apatites, **Table 1** and **Figure 7**.

This figure confirms the less stability of Sr compounds and the particular behavior of the Ca ones.

X in Ca ap.	$\Delta_f G^\circ$ (T°)	x in Sr ap.	$\Delta_f G^\circ$(T°)	x in Ba ap.	$\Delta_f G^\circ$(T°)
0	−12488.276	0	−12438.89	0	−12412.3
0.212	−12601.77	0.16	−12475.81	0.107	−12466.59
0.394	−12673.78	0.35	−12519.21	0.184	−12498.41
0.474	−12692.73	0.45	−12540.16	0.29	−12537.66
0.587	−12706.88	0.47	−12544.15	0.497	−12611.9
0.701	−12718.97	0.49	−12547.14	0.744	−12687.13
0.84	−12678.85	0.51	−12552.13	0.851	−12711.42
0.913	−12688.46	0.56	−12562.61	0.902	−12724.94
1	−12670.77	0.61	−12573.08	1	−12753.79
		0.63	−12577.07		
		0.66	−12581.56		
		0.76	−12598.51		
		0.84	−12612.47		
		1	−12637.38		

Table 1. Estimated free energies of formation of the A-carbonate apatites.

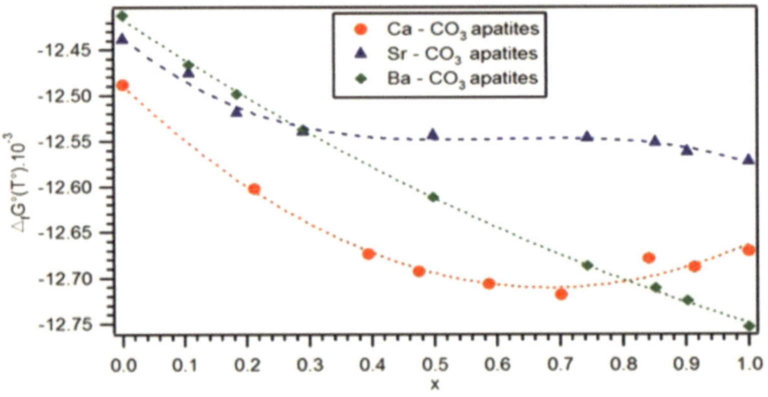

Figure 7. Variation of the standard Gibbs free energies of formation of the type A-carbonate F-apatites as functions of carbonate content 'x'.

4. Thermodynamics of some B-type carbonate Ca-fluorapatites

4.1. Formation enthalpy

Beside the procedure developed previously for determining the formation enthalpy of an apatite, this quantity can also be obtained from dissolving the apatite then other compounds whose dissolutions lead to the same entities in solution as that obtained with the apatite.

http://dx.doi.org/10.5772/68087

However, dissolution of the later introduces new entities in the thermochemical cycle. These entities are compensated by considering supplementary reactions involving dissolution or formation of well-known products. A typical case is illustrated by the cycle leading to the formation enthalpy of the apatite of formula: $Ca_{10-x+u}(PO_4)_{6-x}(CO_3)_xF_{2-x+2u}$ (u instead of y in Section 2) [17]:

$$\{Ca_{10-x+u}(PO_4)_{6-x}(CO_3)_xF_{2-x+2u}\}_{sol} \rightarrow (Ca_{10-x+u}(PO_4)_{6-x}(CO_3)_xF_{2-x+2u})_{(sd)} + \text{solvent} \quad (5)$$

$$x\,CaCO_{3(sd)} + \text{solvent} \rightarrow \{x\,CaCO_3\}_{sol} \quad (6)$$

$$x\,Ca + x\,C + (3/2)x\,O_2 \rightarrow x\,CaCO_{3(sd)} \quad (7)$$

$$(2-x+2u)\,NaF_{(sd)} + \text{solvent} \rightarrow \{(2-x+2u)\,NaF\}_{sol} \quad (8)$$

$$(2-x+2u)\,Na + \left(\frac{2-x+2u}{2}\right)F_2 \rightarrow (2-x+2u)\,NaF_{(sd)} \quad (9)$$

$$\left\{\left(\frac{2-x+2u}{3}\right)Na_2HPO_4.2H_2O_{(sd)}\right\}_{sol} \rightarrow \left(\frac{2-x+2u}{3}\right)Na_2HPO_4.2H_2O_{(sd)} + \text{solvent} \quad (10)$$

$$\left(\frac{2-x+2u}{3}\right)Na_2HPO_4.2H_2O_{(sd)} \rightarrow \left(\frac{4-2x+4u}{3}\right)Na + \left(\frac{2-x+2u}{3}\right)P + \left(\frac{5*(2-x+2u)}{6}\right)H_2 + (2-x+2u)O_2 \quad (11)$$

$$\left(\frac{10-2x+u}{3}\right)Ca_3(PO_4)_{2(sd)} + \text{solvent} \rightarrow \left\{\left(\frac{10-2x+u}{3}\right)Ca_3(PO_4)_{2(sd)}\right\}_{sol} \quad (12)$$

$$(10-2x+u)\,Ca + \left(\frac{20-4x+2u}{3}\right)P + \left(\frac{40-8x+4u}{3}\right)O_2 \rightarrow \left(\frac{10-2x+u}{3}\right)Ca_3(PO_4)_{2(sd)} \quad (13)$$

$$\left\{\left(\frac{2-x+2u}{3}\right)NaOH_{(sd)}\right\}_{sol} \rightarrow \left(\frac{2-x+2u}{3}\right)NaOH_{(sd)} + \text{solvent} \quad (14)$$

$$\left(\frac{2-x+2u}{3}\right)NaOH_{(sd)} \rightarrow \left(\frac{2-x+2u}{3}\right)Na + \left(\frac{2-x+2u}{6}\right)O_2 + \left(\frac{2-x+2u}{6}\right)H_2 \quad (15)$$

$$(2-x+2u)H_2O_{(lq)} + \text{solvent} \rightarrow \left\{(2-x+2u)H_2O_{(lq)}\right\}_{sol} \quad (16)$$

$$(2-x+2u)H_2 + \left(\frac{2-x+2u}{2}\right)O_2 \rightarrow (2-x+2u)H_2O_{(lq)} \quad (17)$$

$$(10-x+u)Ca + (6-x)P + xC + \frac{(2-x+2u)}{2}F_2 + \frac{24-x}{2}O_2 \rightarrow (Ca_{10-x+u}(PO_4)_{6-x}(CO_3)_xF_{2-x+2u})_{(sd)} \quad (18)$$

With $0 \le x \le 2$ and $u \le x/2$ and the subscript "sol" means "in solution."

This cycle gathers a succession of 13 steps whose "summation" leads to step 14 which represents the formation reaction of the apatite. Steps 1, 2, 4, 6, 8, 10, and 12 are dissolution reactions (or their reverse) of the carbonate apatite and well-known products. Their enthalpies were determined by performing experiments at 298 K in the same solvent (aqueous solution having 26.22 wt% H_3PO_4) using a SETARAM microcalorimeter. Steps 3, 5, 7, 9, 11, and 13 are formation reactions or their reverse. The corresponding enthalpies were picked from the literature.

4.2. Formation of Gibbs free energy

Calculation of the formation entropy has been undertaken according to the procedure developed above considering the negative entropy value of the following reaction between CaO, $Ca_3(PO_4)_2$, and CaF_2 as:

$$\mathrm{x\,CaCO_{3(sd)}} + \frac{6-x}{2}\,\mathrm{Ca_3(PO_4)_{2(sd)}} + \frac{2-x+2u}{2}\,\mathrm{CaF_{2(sd)}} \rightarrow \mathrm{Ca_{10-x+u}(PO_4)_{6-x}(CO_3)_xF_{2-x+2u\,(sd)}}$$

with $0 \leq x \leq 2$ and $u \leq x/2$.

This led to derive the formation entropy of the carbonate apatite as a function of x and u and to determine the Gibbs free energy of formation for free carbonate FAP and four carbonate F-apatites having x in the range 0.27–1.22, **Table 2** [17]. One can notice an increase in the later quantity as the carbonate content increases. It seems that, in opposition of the A-type case, introduction of carbonate to form B-type carbonate apatites results in increasing the Gibbs energy and thus reducing the stability of the edifice.

Chemical formula			$\Delta_fH°$/kJ mol^{-1}	$\Delta_fS°$/J mol^{-1} K^{-1}	$\Delta_fG°$/kJ mol^{-1}	S°/J mol^{-1} K^{-1}
$Ca_{10}(PO_4)_6F_2$	–	–	−13744	–	−12983	775.7
	–	–	−13548	−2423	−12825	–
	–	–	−13744	–	−12979	775.8
$Ca_{9.88}(PO_4)_{5.73}(CO_3)_{0.27}F_{2.03}$			−13511	−2514	−12762	770.3
$Ca_{9.84}(PO_4)_{5.49}(CO_3)_{0.51}F_{2.20}$			−13390	−2495	−12646	770.5
$Ca_{9.65}(PO_4)_{5.02}(CO_3)_{0.98}F_{2.27}$			−13033	−2438	−12306	760.2
$Ca_{9.55}(PO_4)_{4.78}(CO_3)_{1.22}F_{2.32}$			−12852	−2408	−12134	755.3

[a] = values from the literature.

Table 2. Standard molar enthalpies, entropies, and Gibbs free energies of formation of the "B" type carbonate fluorapatites at the temperature T = 298.15 K and pressure p = 0.1 MPa [17].

5. Thermochemistry of B-type carbonate Ca/Mg fluorapatites

In natural phosphates, calcium is weekly substituted by other cations, among them magnesium is the most important. Even existing in a few amounts, this element influences the thermochemical properties of the apatite and its reactivity toward the acid attack.

In contrast with other elements such as strontium, barium, lead, or cadmium, substitution of calcium by magnesium in fluor and hydroxyl-apatites does not exceed one atom per unit cell [18], while substitution by the other elements can lead to continuous solid solutions.

A series of B-type carbonate Ca/Mg fluorapatites containing various amounts of carbonate and 0.90 ± 0.01 Mg atom per unit cell have been prepared, characterized, and chemically analyzed. They were then dissolved in a 3 wt% HCl solution [or HCL, 53.53 H_2O] using the isoperibol calorimeter. During dissolution, the solution was saturated by a continuous bubbling of CO_2 gas. Many other complementary experiments were performed in the same device and solvent in order to get the formation enthalpies of the synthesized products. These processes concern dissolution of solids $CaCl_2$, $2H_2O$, and $MgCl_2$ and dilution of [H_3PO_4, 0.756 H_2O], [HF, 1.708 H_2O]. The results were combined with the required formation enthalpies picked from the literature to give the formation enthalpies represented in **Figure 8** together with that corresponding to free magnesium carbonates. The curve corresponding to the latter was drawn from values reported in reference [17].

One can notice that the introduction of 0.90 ± 0.01 Mg atom in the unit cell results in the increase of the formation enthalpy and so a decrease in stability. This is in agreement with the results reported by Drouet [19] in the compilation of thermodynamic quantities of a large number of apatites. Besides, the difference in stability depends on the carbonate content in the lattice, the higher the latter the less stable is the Mg-bearing apatite compared to its homologous Mg-free one.

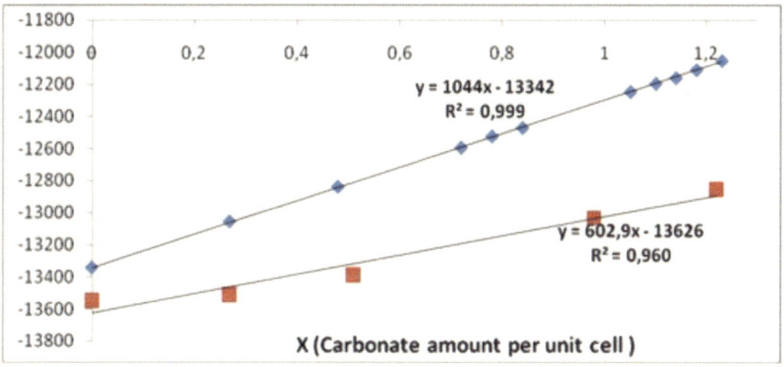

Figure 8. Standard enthalpies of formation of B-type carbonate Ca and Ca/Mg fluorapatites (red color for Ca [17] and blue for Ca/Mg).

6. Thermochemistry of some B-type carbonate hydroxyapatites

B-carbonate Ca-hydroxyapatites have been synthesized, characterized, and chemically analyzed. They were then dissolved in a 9 wt% nitric acid solution (or H_3PO_4, 35.35 H_2O) using the isoperibol calorimeter. The dissolution reaction of the compound having the general formula $Ca_{10-(a+b)}(PO_4)_{6-(a+2b+e)}(CO_3)_{(a+2b+e)}F_{2-(a-e)}$ has been engaged in a succession of other reactions involving dissolution of $Ca(NO_3)_2$,$4H_2O$, dilution of H_3PO_4, 35.35 H_2O, and formation of well-known products and entities [20]. Each of these processes has been affected by a suitable factor so as the "summation" of the reaction succession led to the formation enthalpy of the apatite. The results allowed to derive the latter quantity as a function of the dissolution enthalpy [$\Delta_{sol}H°$(CO3Ap)] and "a," "b," and "c" factors as:

$$\Delta_f H°(\text{apatite}) = -\Delta_{sol}H°(\text{CO3Ap}) - 5144.9 - 1113.74.(10 - a - b) - 1279.7.(6 - a - 2b - e) - 965.16.(a + 2b + e)$$

In which

$$(10 - a - b) = nCa,\ (6 - a - 2b - e) = nPO_4,\ \text{and}\ (a + 2b + e) = nCO_3$$

The mole numbers nCa, nPO_4, and nCO_3 were determined experimentally from chemical analysis. **Table 3** gathers values of standard molar enthalpies of solution and formation for the group of samples.

As with the fluorine homologous compounds, the formation enthalpy increases with the amount of carbonate and thus the stability decreases.

It is assumed that only two kinds of carbonates are present in such compounds, consequently, among *"a," "b,"* and *"e"* parameters, one has to be nil. Experimental results have been treated statistically in order to find which couple of parameters could be retained. Three relationships were derived from the general expression of the formation enthalpy and a mathematical model based on linear regression on two parameters was applied. The couple of parameters considered were *(a,b)*, *(a,e)*, and *(b,e)*. A *F*-test analysis of variance was performed in order to get the significance level of values of the coefficients. The results show that *"a"* and *"b"* coefficients seem to be more significant in the expression of the formation enthalpy. So among I, II, and V mechanisms, the first two are the most probable ones. It seems that substitution of CO_3 for PO_4 in hydroxyapatites is more probably accompanied by the appearance of Ca an OH vacancies according to the two following mechanisms:

	$\Delta_{sol}H$ (kJ mol^{-1})	$\Delta_f H$ (kJ mol^{-1})
$Ca_{9.83}(PO_4)_{5.82}(CO_3)_{0.28}(OH)_{1.64}$	−387.76	−13133.3
$Ca_{9.60}(PO_4)_{5.55}(CO3)_{0.53}(OH)_{1.49}$	−397.44	−12763.3
$Ca_{9.35}(PO_4)_{5.34}(CO_3)_{0.67}(OH)_{1.34}$	−396.05	−12352.6
$Ca_{9.06}(PO_4)_{4.95}(CO_3)_{0.83}(OH)_{1.61}$	−440.99	−11640.0
$Ca_{8.93}(PO_4)_{4.83}(CO_3)_{1.06}(OH)_{1.25}$	−421.71	−11583.0
$Ca_{8.82}(PO_4)_{4.74}(CO_3)_{1.31}(OH)_{0.80}$	−389.58	−11618.7
$Ca_{8.86}PO_4)_{4.65}(CO_3)_{1.37}(OH)_{1.03}$	−420.20	−11575.4
$Ca_{8.76}(PO_4)_{4.59}(CO_3)_{1.44}(OH)_{0.87}$	−400.91	−11474.1
$Ca_{8.68}(PO_4)_{4.52}(CO_3)_{1.58}(OH)_{0.64}$	−387.48	−11443.9
$Ca_{8.51}(PO_4)_{4.41}(CO_3)_{1.63}(OH_{0.53}$	−401.27	−11148.3

Table 3. Molar enthalpies of solution and formation for the B-type carbonate hydroxyapatites.

http://dx.doi.org/10.5772/68087

$Ca^{2+} + PO_4^{3-} \leftrightarrow V^{Ca} + CO_3^{2-} + V^{OH}$ and $Ca^{2+} + 2PO_4^{3-} \leftrightarrow V^{Ca} + 2CO_3^{2-}$ (and not mechanism V as reported erroneously in the conclusion and abstract of reference [21]).

7. Reactivity of B-carbonate fluorapatites toward the acid attack

7.1. Simple dissolution of Ca-CO_3apatites

One of the criterions characterizing a phosphate ore is the speed of its reaction with acid solutions. This influences the reaction yield at a certain time and the self-heating rate of the reactional medium. The acid attack of synthetic and natural phosphates has been largely reported in the literature, but the quasitotality of works was performed by analysis of samples taken from the reactional medium. This way of doing does not give a true image of what is happening during the reaction. Microcalorimetry can overcome this drawback. However, this requires some preliminary experiments and signal treatment.

A C-80 SETARAM microcalorimeter has been adjusted in order to get calibration experiments in the same conditions as the chemical processes. This has been done by supplying the reversal cells of the device with electrical resistances that have been connected to a DC power supplier. When electrical energy is injected during a certain time in the reaction cell, this corresponds to an energy supplied in a "rectangular" shape but leads to a delayed and deformed recorded signal. This is because of the inertia of the various components of the device. A mathematical treatment allows to determine the time constants of the device and to get back the "rectangular" shape of the signal [21]. This "deconvolution" operation is particularly needed in studying fast phenomena. The time constants determined in the electrical calibration operation are then used for the chemical process in order to get a "deconvoluted" calculated curve which represent more accurately what happened inside the reaction cell when the reactants were mixed.

A series of B-type carbonate apatites having 1.53–5.94 wt% carbonate have been synthesized and characterized and then dissolved in 19w/w P_2O_5 solution between 25 and 55°C using the C80- microcalorimeter. Dissolution of 10–60 mg of solid in 4.5 ml acid solution is a simple phenomenon without any resulting precipitate. The drawing of the energy released as a function of the solid amount led to the molar dissolution enthalpies as -201 ± 4, -208 ± 5, and -205 ± 5 kJ mol^{-1} for the 1.05, 3.05, and 5.94 wt% carbonate, respectively [22]. These values are lower than that corresponding to dissolution of carbonate-free fluorapatite in the same conditions (−171 kJ mol^{-1}). Introduction of B-carbonate in the lattice decreases the dissolution enthalpy and so confirms the decrease in stability.

As the dissolution is very fast, the deconvoluted curves were calculated taking into account the time constant values determined in calibration experiments. They were then analyzed according to Avrami model. This model links the reactant transformed fraction "x" to time through the following relationship:

$$-\ln(1 - x) = kt^n$$

With k and n as the Avrami constants and "x" equals the ratio of the heat q released at time t over the overall heat Q_t determined by integrating the whole peak. Originally, this model was developed to interpret the kinetics of crystallization of a compound from its molten state. The procedure consists in drawing ln[−ln(1 − X)] as a function of lnt. The existence of a slope change in the curve has been attributed to the appearance of a new phenomenon during a phase change. The model was then applied to the dissolution and precipitation process, such as dissolution of metals and oxides. Other studies also used this model for the interpretation of the dissolution and precipitation results of several products.

Applying this model to B-carbonate apatites reveals the existence of two processes, **Figure 9**, with a very short duration of the first one (33 seconds). It has been attributed to a surface phenomenon. The second one that lasts till 1200 seconds has been exploited to get a kinetic model of the acid attack. The results seem to be in agreement with a homogeneous model having an order of 2 with respect to the apatite. Activation energy values were deduced as 15.8 ± 1.4, 15.5 ± 0.9, and 8.5 ± 0.5 kJ mol^{-1} for the 1.58, 3.05, and 5.94 wt% carbonate, respectively [22]. These values are of the same order of magnitude as that determined for pure FAP below 45°C (16 kJ mol^{-1}) but drastically lower than that determined at higher temperature for the latter compound (101 kJ mol^{-1}) [23]. The lower value of the activation energy indicated that the reaction is rather controlled by a diffusion phenomenon.

7.2. Simple dissolution of Ca/Mg-CO_3apatites

Different masses of Ca/Mg bearing B-type carbonate apatites having various amounts of carbonate have been dissolved in 3 wt% HCl solution at 25–55°C using the C-80 microcalorimeter. For low values of the solid/liquid ratio, no solid resulted and the recorded signals show the appearance of a process lasting until 2000 seconds. The drawing of the energy released at 25°C as a function of the solid mass allows to deduce the molar dissolution enthalpies from the slopes of the lines, **Table 4**.

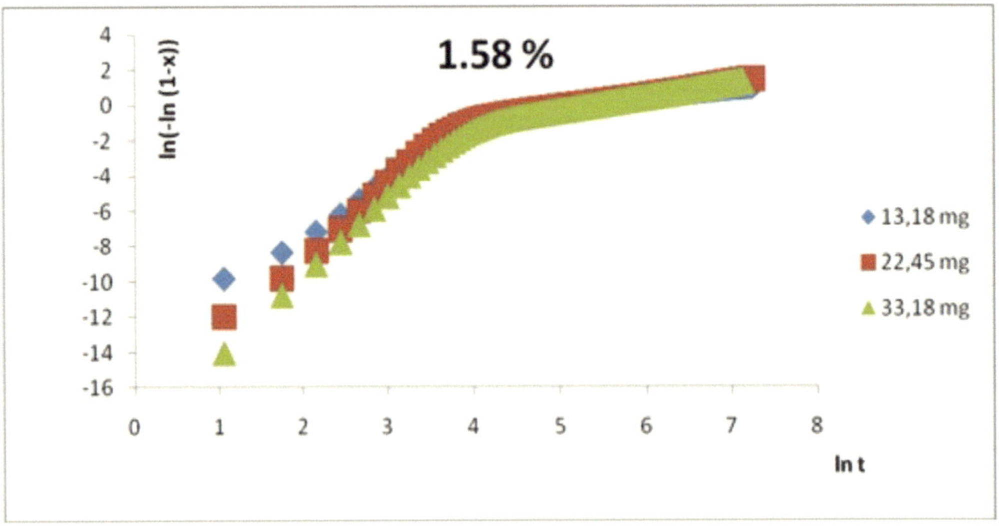

Figure 9. Example of Avrami curves recorder for various masses of 1.58 w% CO_3 B-type F-apatite.

http://dx.doi.org/10.5772/68087

Apatite	$\Delta_{sol}H°(T°)$	Apatite	$\Delta_{sol}H°(T°)$
$Ca_{10}(PO_4)_6F_2$	−151.7 ± 0.7	$Ca_{8.60}Mg_{0.09}(PO_4)_{5\ .16}(CO_3)_{0.84}F_{1.83}$	−209.7 ± 1.1
$Ca_{9.09}Mg_{0.91}(PO_4)_6F_2$	−180.6 ± 0.1	$Ca_{8.36}Mg_{0.09}(PO_4)_{4.77}(CO_3)_{1.23}F_{1.75}$	−222.7 ± 0.3
$a_{9.09}Mg_{0.09}(PO_4)_{5\ .52}(CO_3)_{0.48}F_{1.90}$	−192.3 ± 05		

Table 4. Molar dissolution enthalpy in 3wt% HCl solution for Ca-FAP and Ca/Mg fluorapatites containing 0.90 ± 0.01 Mg and various amounts of carbonate.

These values are higher than that obtained by dissolving the same products in the isoperibol calorimeter [24] and their difference increases with the carbonate amount. This is because the latter experiments were performed in a CO_2 saturated solution, and so the solution is supposed to retain any further CO_2, while in the microcalorimeter the dissolution occurs in a hermetic cell and the CO_2 released is partially dissolved in the resulting solution. This results in an exothermal increment in the dissolution energy that increases with the carbonate amount in the solid apatite. Appearance of such effect is in agreement with the diminution of solubility of gases in liquids when increasing temperature.

One can notice a decrease in the dissolution enthalpy when Mg is incorporated and a further decrease when introducing carbonates. Introduction of magnesium and carbonate deceases considerably the molar dissolution enthalpy.

The recorded signals were processed in order to get the corresponding deconvoluted (or thermogenesis) curves [25]. As previously, the Avrami curves drawn from the latter show at first very short phenomenon followed by a longer one. Treatment of the data reveals as previously, that the experimental results are in agreement with a homogeneous kinetic scheme with an order of 2 with respect to the apatite and an activation energy, E_a, decreasing from 20.1 ± 0.7 to 18.6 ± 0.1 kJ mol^{-1} when only magnesium is introduced in the lattice. E_a decreases again to 6.0 ± 08 kJ mol^{-1} when the carbonate amount in the Mg-bearing apatites reaches 1.23 mol CO_3 in the formula unit. The presence of magnesium and carbonate in B-carbonate F-apatites also seems to significantly decrease the activation energy of the acid attack and hence to increase considerably the velocity of the reaction.

However, the homogeneous kinetic model is far from the experimental reality since the reaction occurs between a solid and a liquid. Isoconversional model is more suitable for heterogeneous processes. According to this model, the converted fraction α of a reactant is expressed as a function of time by the equation:

$$\frac{d\alpha}{dt} = kf(\alpha)$$

with k the rate constant and $f(\alpha)$ a function associated to the mechanism. Integration of this equation leads to:

$$g(\alpha) = \int_0^\alpha \frac{dx}{f(\alpha)} = kt$$

Considering the Arrhenius law, $g(\alpha)$ can be expressed as: $g(\alpha) = A\exp\left(-\frac{E_a}{RT}\right)t$

and so $\ln(t) = \frac{E_a}{RT} + \ln\left(\frac{g(\alpha)}{A}\right)$

At a certain conversion rate, $g(\alpha)/A$ is constant and so it is possible to determine the activation energy, whatever is the mechanism, by plotting ln(t) versus $1/T$. Calculation was performed for α in the range 0.06–0.98, and the activation energy was derived.

Examples of plots of ln(t) versus $1/T$ ($25 \leq T \leq 55°C$) for the Ca/Mg apatite containing 0.48 CO_3 are given in **Figure 10** (similar curves were obtained for all the apatites). The variation of the activation energy as a function of α is represented in **Figure 11** for all the apatites with the corresponding errors calculated by considering the scatter of the points around the least square line drawn for each value of α in **Figure 10** [25].

One can notice that the activation energy of any apatite is practically constant along all the process, suggesting a single-step reaction. Statistical calculations allowed to determine this quantity together with its error. The values are in the range 21.1 ± 0.3–6.2 ± 0.2 kJ mol^{-1} for the series of compounds. It should be noticed that these values of E_a are in the range of their corresponding previous ones determined from the Arrhenius plots in the homogeneous kinetic model.

7.3. Dissolution of more amounts of Ca-CO_3apatites

As with pure fluorapatite reported in Ref. [7], progressive dissolution of carbonated Ca-apatites in the same amount of a 19% w/w P_2O_5 solution (4.5 ml) results in the occurrence of successive phenomena. This appears through the graph representing the energy released as a function of the solid amount. **Figure 12** shows an example corresponding to dissolution of the

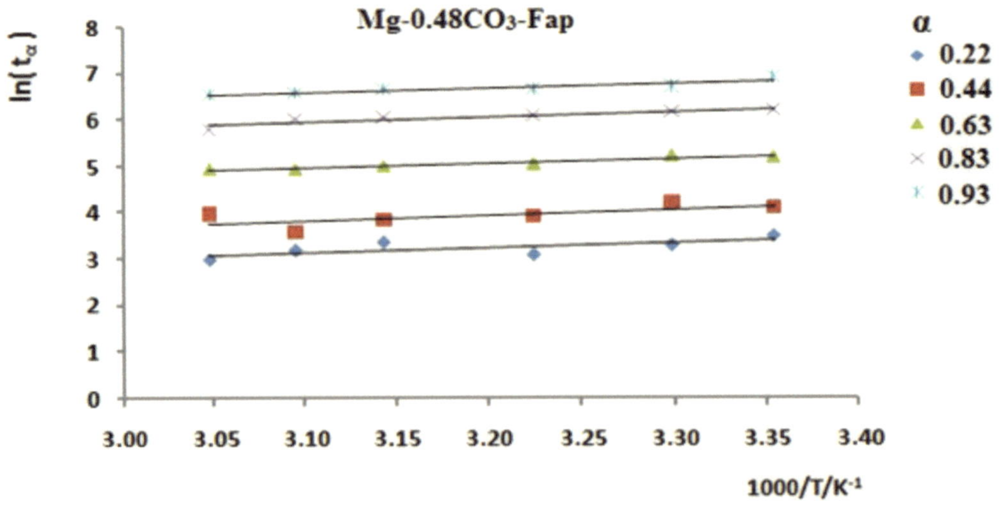

Figure 10. ln(t) *versus* $1/T$ ($25 \leq T \leq 55°C$) for the 0.48 CO_3 in Ca/Mg F-apatite.

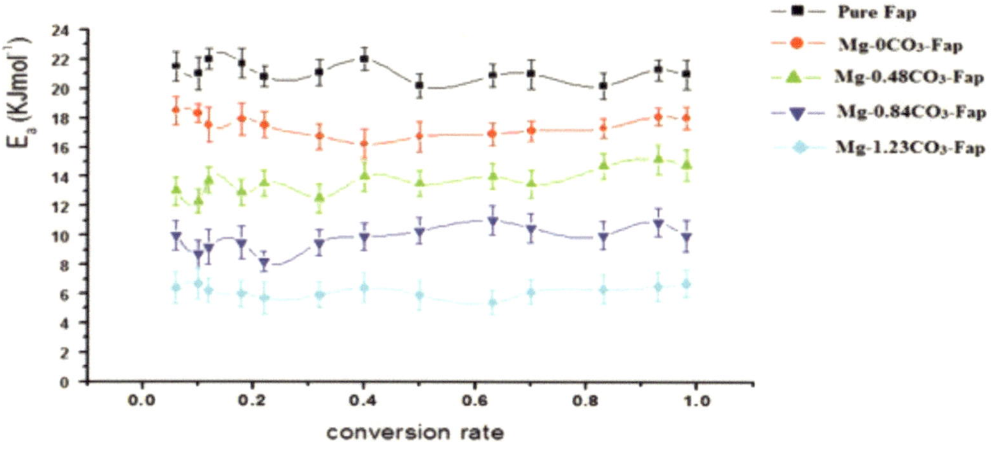

Figure 11. Activation energies, E_a, *versus* the conversion rate, α, for Ca/Mg F-apatites containing 0–I.23 carbonate in their lattice.

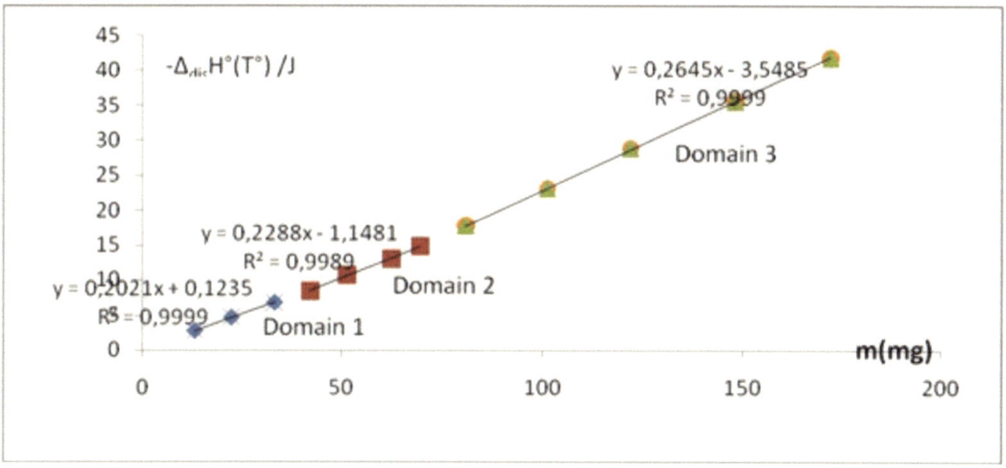

Figure 12. Heat energy at 25°C *versus* the mass of the 1.58 B-type carbonate F-apatite.

1.58 carbonate sample versus the solid mass. One can notice the presence of three domains, each one corresponding to a particular process.

Domain 1 corresponds to a simple dissolution, no precipitate appeared at the end of the process, while in domain 2 dissolution is followed by precipitation of CaF_2 and in domain 3 precipitation of a mixture of the latter with MCPM (mono-calcium phosphate monohydrate, $Ca(H_2PO_4)_2,H_2O$). These precipitates were characterized by X-ray diffraction performed on the solids after reaction [26]. All the raw thermograms recorded in these domains indicated a unique phenomenon, while the thermogenesis curves in domains 2 and 3 show the successive appearance of two or three phenomena [26].

8. What about the phosphate ore?

It would be interesting to compare the behavior of the phosphate ore to that of these products. However, as the latter contains many compounds, dissolving a few dozens of milligrams is not representative of what is really happening. The differential reaction calorimeter (DRC) allows to overcome this difficulty. This device is composed of two symmetrical glass reactors of about half a liter maintained at the same temperature by a fluid circulating in a double envelop. Two axial agitators running at the same speed allow to disperse the solid in the liquid bulk. The former is initially introduced in a tubular sample holder and tightly separated from the liquid by a thin film blocking up the lower extremity of the tube. The recorded signal is a DC voltage resulting from the temperature difference between the two reactors that is detected by two thermocouples whose junctions are immersed in the liquids. The device has been calibrated and tested by experiments performed on key reactions [27].

A sample ore from Gafsa, Tunisia, has been previously chemically and structurally analyzed. It contains about 5.7 w% $CaCO_3$ and 82.4 wt% of an apatite phase of formula: $Ca_{9.25}Na_{0.52}Mg_{0.23}(PO_4)_{5.14}(CO_3)_{0.86}F_{2.34}$. Various amounts (up to 40 g) of solid have been dissolved at 25–65°C in 100 g of an acidic solution having 20, 25, or 30 wt% P_2O_5. These compositions are of the same order of magnitude as that used in industrial manufacturing process of superphosphate fertilizer. As with B-type CO_3 apatites at 25°C, attack of progressive amounts of solid by the same quantity of liquid show three domains each one corresponding to a particular phenomenon, **Figure 13**. Similar curves were obtained with 25 and 30 wt% P_2O_5.

The first domain corresponds to a simple dissolution with $\Delta_{diss}H°(T°)$ in the range -189.7 ± 2.1 to -138.9 ± 5.5 J.g^{-1} for 20, 25, and 30 wt% P_2O_5. The first value lies between that determined for the free Mg B-carbonate fluorapatites in 19 wt% P_2O_5 (around -0.824×212 to -172.5 J g^{-1} [17]) and that determined for Mg/$CaCO_3$-apatite containing 0.84 CO_3 in the lattice (around -0.824×250 to -206 J g^{-1} [24]). The scatter cannot be explained by the presence of calcite in the ore, it could result from the presence of sodium in the apatite ore. Dissolution of Ca/Na carbonate F-apatites will help to get a better explanation. The second domain corresponds to dissolution then precipitation

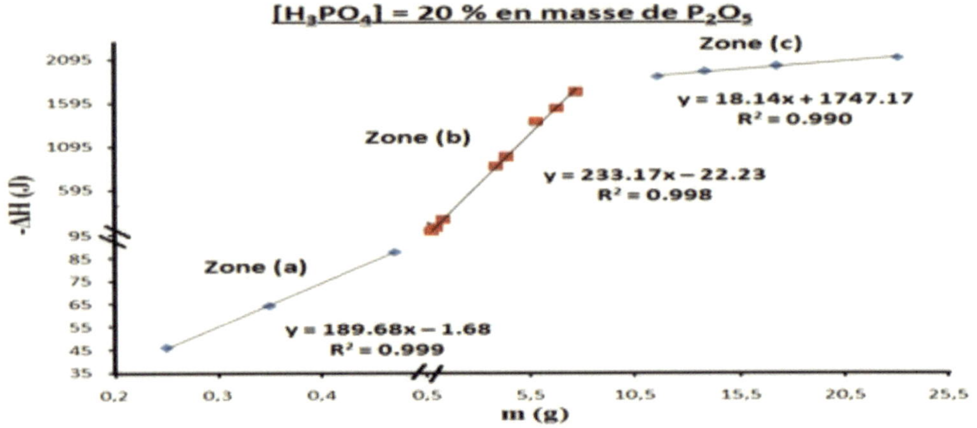

Figure 13. Heat energy released at 25°C by dissolving m(g) of the Gafsa ore in 100g acid solution having 20wt% P_2O_5.

http://dx.doi.org/10.5772/68087

of MCPM, $Ca(H_2PO_4)_2$, H_2O (and not CaF_2 as with synthetic B-carbonates), and the third one to the appearance of a mixture of the latter with di-calcium phosphate dihydrate, DCPD, $CaHPO_4$,$2H_2O$ [28]. **Figure 13** also allows to determine the solubility of MCPM and DCPD in the acid solution. These quantities are calculated considering the intersection point of the straight lines and equal 0.47, 0.52 and 0.62 g for MCPM and 8.4, 11.8, and 10.3 g of solid per 100 g solvent for DCPD at 25°C in the three acid concentrations, respectively. At 65°C only domains 2 and 3 appeared, suggesting a very week solubility of MCPM in these solutions at that temperature.

The kinetic model of the attack in domain 3 has established considering the thermogenesis curves. The Shrinking Core Model (SCM) seems to be in agreement with experimental results. According to this model, the reaction occurring between a solid and a liquid is controlled by one of the main following processes:

(1) diffusion of the liquid reactant through a liquid film surrounding the particle of the solid to the surface of the latter, (2) chemical reaction at the surface of the nonreacting solid, leading to the reaction product (or ash), and (3) diffusion of the resulting liquid through the ash back to the external surface of the solid. The following equations have been proposed for these processes [29]: $X = kt$, $[1 - (1 - X)^{1/3}] = kt$ and $[1 - 3(1 - X)^{2/3} + 2(1 - X)] = kt$, where X is the conversion fraction of solid, k is the kinetic constant, and t is the time.

Calculation performed on the thermogenesis curves lead to conclude that the experiment results are in agreement with the third process. An activation energy value has been deduced as 25.4 ± 1.8 kJ mol^{-1}, and so the acid attack seems to be controlled by the ash diffusion phenomenon.

However, this model describes globally the process and does not take into account the evolution of the reaction scheme resulting from the modification of the grain surface during the reaction. The isoconversional model allows to take into account this phenomenon. Applying this model to the process occurring in domain 3 led to straight lines representing $\ln(t)$ as functions of $1/T$ for various values of the conversion rate, as in **Figure 10**. The deduced activation energy E_a values allow to draw the latter quantity as a function of X, **Figure 14**.

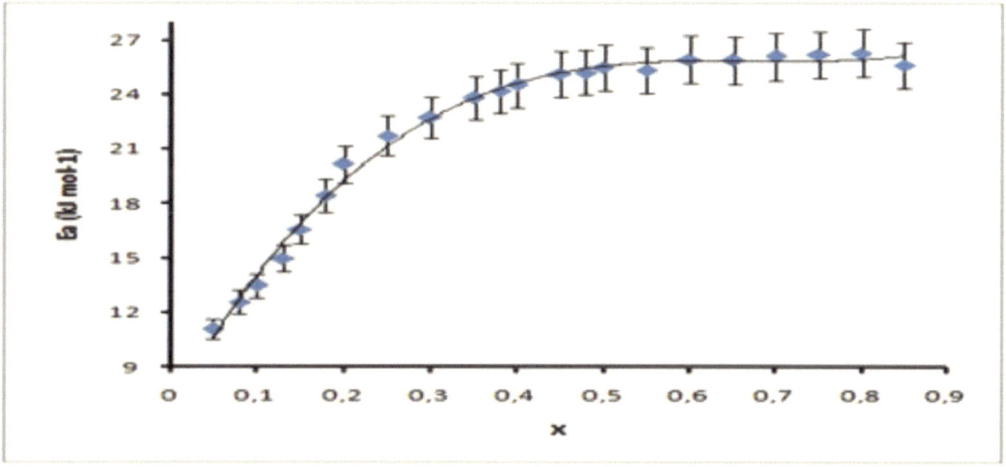

Figure 14. Activation enegy (E_a) as a function of the conversion rate (x) according to the isoconversional model.

Value of E_a determined in the SCM model (25.4 ± 1.8 kJ mol^{-1}) belongs to the interval values corresponding to the isoconversional model, which lies in the range 11.1–26.3 kJ mol^{-1} and the shape of the curve indicates that the process does not occur in only one step. It could be a successive or consecutive reaction process [30].

In conclusion, thermodynamic determinations allow to point out the influence of the various components on the stability and reactivity of the apatites. The results show that introduction of carbonate along the channel contributes to increase the stability of F-apatites, while in B-type fluor- and hydroxyl-apatites, the carbonate ions contribute to decrease the stability. It is also for the substitution of calcium by magnesium. B-CO_3 and Mg-substitutions increase the reactivity of synthetic F-apatites toward the acid attack that seems to be a one step process, while the acid attack of natural phosphate is more complicated.

Acknowledgements

The author would like to thank his collaborators F. Bel Hadi Yahia, K. Brahim, H. Chihi, S. Jebri, I. Khattech, A. Soussi-Baatout, and H. Zendah who contributed in the papers reported under their names in the reference list.

Author details

Mohamed Jemal

Address all correspondence to: jemal@planet.tn

Tunis El Manar University. Faculty of Science, Tunis El Manar, Tunisia

References

[1] Becker P. Phosphates and Phosphoric Acid, Raw Materials, Technology and Economics of the Wet Process. 2nd ed. New York, Basel: Marcel Dekker Inc; 1989. p. 740

[2] Naray-Szabo S. The structure of apatite $(CaF)Ca_4(PO_4)_3$. Zeitschrift für Kristallographie. 1930;**75**:387

[3] Mehmel M. Zeitschrift für Kristallographie. 1930;**75**:323

[4] Elliott J. Structure and Chemistry of the Apatites and Other Calcium Orthophosphates. Amsterdam, London: Elsevier; 1994. p. 389

[5] Gottschall AJ. Journal of South Africa. 1958;**II**:45

[6] Jemal M. Thermochemistry and relative stability of Apatite Phosphates. Phosphorus Research Bulletin. 2004;**15**:119-124

[7] Jemal M. Thermochemistry and kinetics of the reactions of Apatite Phosphates with acid solutions. In: Mizutani T, editor. Application of Thermodynamics to Biological and Materials Science. Croatia: Intech Open Access Publisher; 2011. pp. 547-572

[8] Lafon JP, Champion C, Bernache-Assollant D, Gibert R, Danna AM. Thermal decomposition of carbonated calcium phosphate apatites. The Journal of Thermal Analysis and Calorimetry. 2004;**72**(3):1127-1134

[9] De Maeyer EAP, Verbeeck RMH. Possible substitution mechanisms for sodium and carbonate in calcium hydroxyapatite. The Bulletin des Societes Chimiques Belges. 1993;**102**:601-609

[10] Kühl G von, Nebergall WH. Hydrogenphosphat- and carbonatapatite. Zeitschrift für anorganische und allgemeine Chemie. 1963;**324**:313-320. Cited in reference 4 page 234

[11] Jebri S, Khattech I, Jemal M. Standard enthalpy, entropy and Gibbs free energy of formation of 'A' type carbonate phosphocalcium Hydroxyapatites. The Journal of Chemical Thermodynamics. 2017;**106**:84-94

[12] Jebri S, Bouzhzala H, Bechrifa A, Jemal M. Structural Analysis and Thermochemistry of 'A' type phosphostrontium carbonate hydroxyapatites. The Journal of Thermal Analysis and Calorimetry. 2012;**107**:963-972

[13] Jebri S, Bechrifa A, Jemal M. Standard enthalpies of formation of 'A' type cabonate phosphobaryium hydroxyapatites. The Journal of Thermal Analysis and Calorimetry. 2012;**109**:1059-1067

[14] Ben Chérifa A, Jemal M. Sur la réaction de dissolution des phosphates dans les acides: Enthalpie de dissolution du phosphate tricacique β dans l'acide nitrique. Annales de Chimie - Science des Materiaux. 1985;**10**:543-548

[15] Stull DR, Prophet H. JANAF Thermochemical Tables. 2nd ed. Washington, DC: NSRDS-NBS 37; 1971

[16] Lide DR. Handbook of Chemistry and Physics. 79th ed. London, NY: CRC Press; 1998-1999

[17] Zendah H, Khattech I. Standard enthalpy, entropy and Gibbs free energy of formation of "B" type carbonate fluorapatites. The Journal of Chemical Thermodynamics. 2015;**87**:29-33

[18] Ben Abdelkader S, Khattech I, Rey C, Jemal M. Synthèse, caractérisation et thermochimie d'apatites calco-magnésiennes hydroxylées et fluorées. Thermochimica Acta. 2001;**276**:25-36

[19] Chihi H, Khattech I, Jemal M. Preparation, characterization and thermochemistry of magnesium co-substituted fluorapatites. The Journal of Thermal Analysis and Calorimetry. 2017;**127(3)**:2427-2438. DOI: 10.1007/s10973-016-5693-2

[20] Drouet C. A comprehensive guide to experimental and predicted thermodynamic properties of phosphate apatite minerals in view of applicative purpose. The Journal of Chemical Thermodynamics. 2015;**81**:143-159

[21] Bel Hadj Yahia F, Jemal M. Synthesis, structural analysis and thermochemistry of B-type carbonate hydroxyapatites. Thermochimica Acta. 2010;**505**:22-32

[22] Brahim K, Khattech K, Dubès JP, Jemal M. Eude cinétique et thermodynamique de la dissolution de la fluorapatritedans l'acide phosphorique. Thermochimica Acta. 2005;**436**: 43-50

[23] Zendah H, Khattech I, Jemal M. Thermochemical and kinetic studies of the acid attack of "B" type carbonate fluorapatites at different temperatures (25–55°C). Thermochimica Acta. 2013;**265**:46-51

[24] Brahim K, Antar K, Khattech I, Jemal M. Effect of temperature on the attack of fluorapatite by a phosphoric acid solution. Scientific Research and Essays. 2008;**3**(1):035-039

[25] Chihi H, Khattech I, Jemal M. Thermochemistry and kinetics of the attack of magnesium-carbonate co-substituted fluorapatites by hydrochloric acid at different temperatures (25–55) °C. Thermochimica Acta. 2016;**646**:16-25

[26] Zendah H, Khattech I, Jemal M. Synthesis, characterization and thermochemistry of acid attack of "B" type carbonate fluorapatites. The Journal of Thermal Analysis and Calorimetry. 2012;**100**:855-861

[27] Soussi-Baatout A, Hichri M, Ben Cherifa A, Khattech I. Test and calibration processes for the differential reaction calorimeter (DRC). Application. Dissolution of calcium fluorapatite in the hydrochloric acid. The Journal of Thermal Analysis and Calorimetry. 2014;**580**:85-92

[28] Soussi-Baatout A. Contribution à l'étude thermochimique et cinétique de l'attaque phosphorique d'un phosphate naturel tunisien [thesis]. Tunisia: Tunis El Manar University; 2016.

[29] Levenspiel O. Chemical Reaction Engineering. 3rd ed. New York, Chichester, Toronto: John Wiley and Sons; 1999. p. 668

[30] Vyazovkin SV, Lesnikovich AI. An approach in the solution of the inverse kinetic problem in the case of complex processes. Thermochimica Acta. 1990;**165**:273-280

Chapter 2

Recent Trends in Phosphatase-Mediated Bioremediation

Gouri Chaudhuri, Uma Selvaraj, P. Venu-Babu and Richard W. Thilagaraj

Additional information is available at the end of the chapter

http://dx.doi.org/10.5772/intechopen.68658

Abstract

Industrial effluents from tanneries and electroplating industries from small- and large-scale sector industrial plants contain substantial amount of toxic heavy metal, which pollutes rivers and lakes, land, air and sea leading to imbalance of ecosystem and certain health issues to humans, animals as well as plants. The worldwide environmental regulations stipulate the reduction of heavy metals in the effluents to permissible levels before discharging into water bodies. Enzyme-mediated precipitation of heavy metals affords a novel eco-friendly method for remediation of toxic heavy metals from various industrial effluents like tannery, electroplating and dye industries. This chapter has paid attention to bacterial alkaline phosphatase (BAP) from *Escherichia coli* C90 and calf-intestinal alkaline phosphatase (CIAP), which catalyses phospho mono- and diesters and produces inorganic phosphate (Pi). The Pi thus generated precipitates the heavy metals as metal-phosphate complexes. The kinetic behaviour of both the enzymes with *para*-nitrophenyl phosphate, ascorbic acid 2-phosphate and α-naphthyl phosphate was investigated at various pH regimes from 8 to 11. The chapter also explains in detail the descriptive information on the capability of BAP- and CIAP-mediated precipitation of heavy metals, which is desirable and convenient method for the toxic heavy metals such as chromium, cadmium, nickel and cobalt.

Keywords: bacterial alkaline phosphatase, calf-intestinal alkaline phosphatase, heavy metals, enzyme kinetics, bioremediation

1. Introduction

Phosphatase-mediated bioremediation of heavy metals plays an important role in the bioremediation of industrial, municipal and nuclear wastewater. Heavy metals, such as Li^{3+}, Mn^{2+}, Cu^{2+}, Zn^{2+}, Cd^{2+}, Ba^{2+}, Hg^{2+}, Pb^{2+}, $Hg2^{+}$, Ag^{2+}, Cd^{2+}, $A1^{3+}$, Ni^{2+}, Cu^{2+} and Pb^{2+}, are found in various industrial effluents. High quantity of Cd^{2+}, Ni^{2+}, Co^{2+} and $Cr^{3+/6+}$ was found in tannery and electroplating effluents which creates high risk in the environment which is one of the important issues in Asian countries. The use of alkaline phosphatase offers great promises in vast research field and could be a model to study metal ion-dependent catalysis to address such a kind of problems in the application field, viz., environmental biotechnology, molecular biology and immunodetection. Alkaline phosphatases are commonly found from bacteria to higher mammals, which efficiently hydrolyze several mono- and diesters. They are nonspecific and react with a variety of substrates from natural to synthetic compounds. There has been a considerable effort in recent years towards the application of alkaline phosphatases for bioremediation of heavy metals and radionuclides from industrial and nuclear wastes. In order to extend bioremediation possibilities to heavy metals contaminated wastes, this present research work explores the advantages of using bacterial (*Escherichia coli* C90) and calf-intestinal alkaline phosphatase enzymes with *p*-nitrophenyl phosphate (*p*NPP), 1-naphthyl phosphate monosodium salt monohydrate and L-ascorbic acid 2-phosphate sesquimagnesium salt hydrate as substrate for enhancing bioprecipitation from single-ion metal solutions (Cd^{2+}, Ni^{2+}, Co^{2+}, Cr^{3+} and Cr^{6+}) and effluents from tannery and electroplating industrial units, which contain Cr^{3+}, Cr^{6+} and Cd^{2+}, Ni^{2+}, Co^{2+}, respectively.

2. Phosphatase enzymes

Phosphatase is a hydrolase enzyme. Phosphatases play a very important role in the phosphate metabolism of the organism by hydrolysis of polyphosphates and organic phosphates [1]. The essential feature of phosphatase is that the enzymes have wide specificity. Phosphatases cleave phosphate ester bonds, which play an important role in the hydrolysis of polyphosphates and organic phosphates. They are also responsible for phosphate transport within the cells [2]. Phosphatase enzymes are used by many soil microorganisms to access organically bound phosphate nutrients.

Two types of phosphatase enzymes are as follows:

- Acid phosphatases
- Alkaline phosphatases

2.1. Alkaline phosphatase

Alkaline phosphatases (APase or AP or ALP; orthophosphoric monoester phosphohydrolase, EC 3.1.3.1) are common in a wide variety of bacteria and mammals which efficiently hydrolyse several mono- and diesters [3]. Alkaline phosphatase, which hydrolyses organic phosphate esters and releases inorganic orthophosphate, is a dimeric metallic enzyme [4]. In bacteria, alkaline phosphatase is present in the periplasmic space. The three-dimensional structure of *E. coli* alkaline phosphatase (**Figure 1**), determined by Kim and Wycoff, reported that each of the two identical subunits contain 449 amino acids [5, 6]. Alkaline phosphatase from *E. coli* has been the focus of several studies dealing with molecular properties, subunit composition and catalytic mechanism; however, much less has been done with mammalian

http://dx.doi.org/10.5772/intechopen.68658

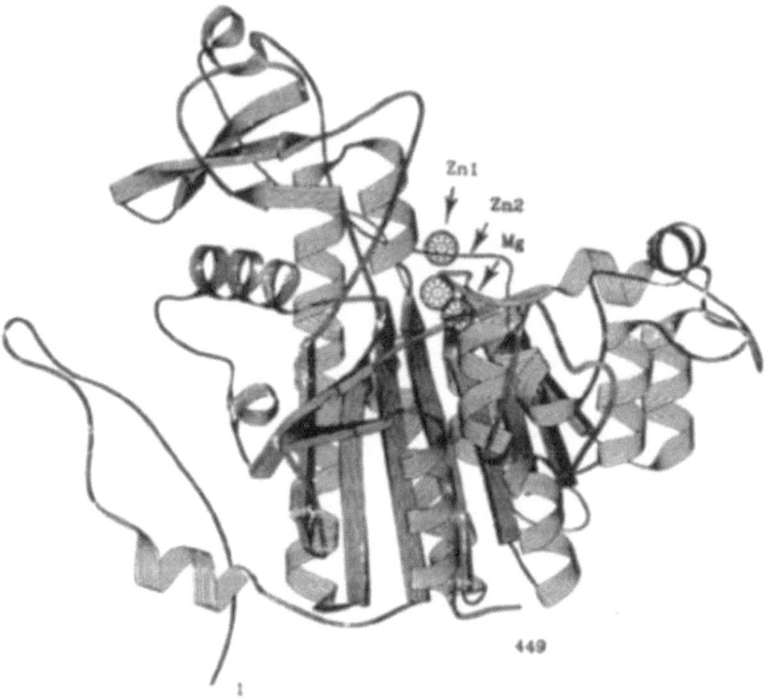

Figure 1. A subunit of *E. coli* alkaline phosphatase shown in ribbon representation. Znl, Zn2 and Mg are indicated as spheres. The monomer shows 10 β-sheets flanked by 15 α-helices of different lengths [12].

alkaline phosphatases. The latter are glycoproteins and exist as different isoenzymes such as placental (PLAP) (**Figure 2**), germ cell (GCAP), intestinal and tissue nonspecific isoforms [3]. The tissue nonspecific isoforms include AP present in bone, liver and kidney and they differ from the other isoforms due to modifications brought about by post-translational events [7]. Mammalian alkaline phosphatase is a very important enzyme physiologically and is an important component of medical diagnosis [8]. ALPs are capable of catalysing the hydrolysis of monoesters of phosphoric acid and also transphosphorylation reaction in the presence of large concentrations of phosphate acceptors. Both bacterial and mammalian ALPs contain two Zn and one Mg on each catalytic site [9] (**Figures 1** and **2**). Both the metal ions are crucial for the enzyme to maintain its structural stability and catalytic activity. About 25–35% similarity was observed in alkaline phosphatase from mammals in the secondary structure and also at the catalytically important residues [9]. The active site residues such as Asp 91, Ser 92, Arg166 and ligands coordinating the divalent metal ions (Zn^{2+} and Mg^{2+}) are all conserved [6]. These structural similarities suggest that mammalian ALPs may catalyse hydrolysis of both pyrophosphate and orthophosphate moieties through a mechanism similar to that of *E. coli* enzyme [10]. Ser-102 residue of the enzyme attacks the phosphoryl group leading to phosphorylated enzyme [11]. The greater advantage of using ALP is its broad substrate specificity.

It can hydrolyse a wide variety of substrates, that is, synthetic as well as natural substrates. The substrates include ATP, ADP, AMP, PPi, glucose-1-phosphate, glucose-6-phosphate,

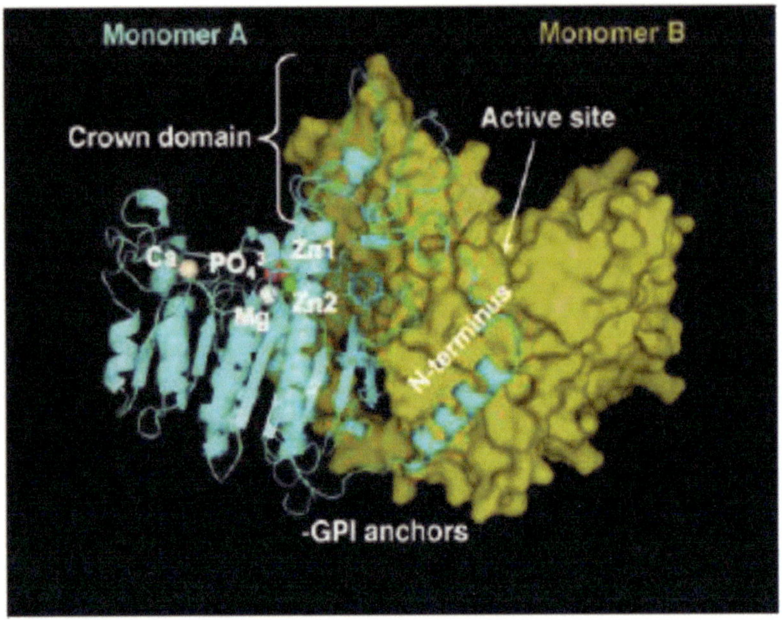

Figure 2. Three-dimensional structure of PLAP. An overview of the structure of human PLAP from the crystallographic coordinates [13]. Ribbon representation of subunit A in cyan and surface representation of subunit B in yellow are shown. Active site metal ions such as Zn1, Zn2 and Mg in addition to the new fourth metal Ca are indicated [9].

fructose-6-phosphate, b-glycerophosphate, bis-(p-nitrophenyl)-phosphate, and so on. [12]. Based on the X-ray crystallography study, a revised mechanism of hydrolysis of a substrate by alkaline phosphatase has been proposed by Stec et al. [14] (**Figure 3**).

This causes a second inversion of configuration at the phosphorus centre. Nucleophilic Ser102 gets regenerated and reprotonation of Ser102 by the Mg-coordinated water molecule may facilitate the departure of the inorganic phosphate product from the non-covalent E.Pi complex. Alternatively, the phosphate group, for its release, may directly get protonated by the Mg-coordinated water molecule [14].

According to this proposed mechanism, the active site of the free enzyme (E) is occupied by three water molecules and the hydroxyl group of S102 forms hydrogen bond with a hydroxide ion which is coordinated by Mg. As a result of the enzyme-substrate complex (E.ROP) formation, the ester oxygen atom coordinates with Zn1 and Zn2 in addition to the guanidinium group of Arg166 [9]. The site opposite to the leaving group gets occupied by Ser102. As a result of binding of the phosphomonoester (ROP) to form the enzyme-substrate complex (E.ROP), the Ser102 O^{γ} becomes fully deprotonated for nucleophilic attack with the consequent transfer of the proton to the Mg-coordinated hydroxide group to form Mg-coordinated water molecule. Coordination of Zn2 stabilizes Ser102 O^{γ} in its nucleophilic state. The activated hydroxyl group of Ser102 attacks the phosphorus centre of the substrate in the enzyme-substrate complex (E-ROP) to form a covalent serine-phosphate intermediate (E-P) resulting in inversion of the phosphorus centre and the loss of the leaving group (RO^-). A nucleophilic hydroxide ion coordinated to Zn1 attacks the covalent serine-phosphate intermediate (E-P), resulting in the formation of non-covalent enzyme-phosphate complex [14].

http://dx.doi.org/10.5772/intechopen.68658

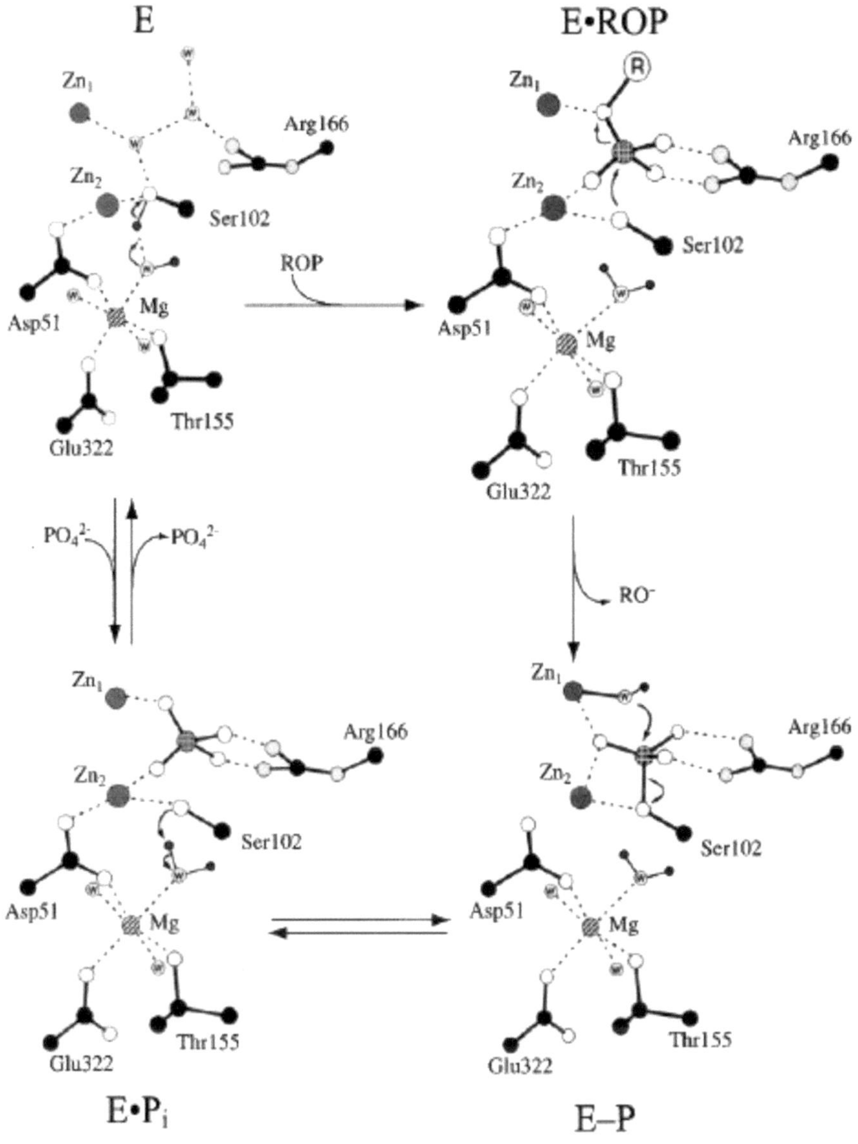

Figure 3. Proposed mechanism of hydrolysis of substrate by alkaline phosphatase [14].

Alkaline phosphatase catalyses the cleavage of phosphate group from *p*-nitrophenyl phosphate and liberates *p*-nitrophenol (PNP) and inorganic phosphate (Pi) [14]. The liberated inorganic phosphate interacts with metal ion or radionuclide and precipitates as metal phosphate or radionuclide phosphate as follows:

$$\text{p-nitrophenyl phosphate} + \text{AP} \rightarrow \text{p-nitrophenol} + \text{phosphate} \quad (1)$$

When reacting with heavy metal:

p-nitrophenyl phosphate + AP + heavy metal → *p*-nitrophenol + heavy-metal phosphate

3. Kinetic behaviour of calf-intestinal and bacterial alkaline phosphatase with pNPP

3.1. Effect of pH and substrate concentration on hydrolysis of CIAP with pNPP

The effects of pH on the kinetics of CIAP-catalysed pNPP hydrolysis reaction were determined in Tris-HCl buffer. CIAP activity was highest at pH 11 in Tris-HCl buffer, whereas in glycine-NaOH buffer the highest activity was recorded at pH 9.5. The rate-determining step at pH 8 was attributed to the non-dissociation of phosphate from the non-covalent enzyme-phosphate complex. At pH 8, the rate of hydrolysis of 4-nitrophenyl phosphate was increased, in the presence of Tris-Cl [15] and also with the substrate 4-methylumbelliferyl [16]. Experimental data show that even at pH 11, Tris buffer was not inhibited. Despite the increase in substrate concentration at pH 11, no reduction in the enzyme activity was observed; by contrast, the activity observed was much higher than that of the activity observed between pH 8.5 and 11 (**Figure 4**). This states that the enzyme exhibits functionally stable conformations at pH 8–8.5 and at pH 11. In the mean time, at a wide substrate concentration between 0.2 and 4 mM the enzyme showed a relatively stable catalytic activity. Similar increase in the pH to 9 was reported previously with ALPs from dog's intestine, kidney and liver [17]. Moreover, at pH 11, the enzyme also showed an optimum temperature at 45°C instead of 37°C. Meanwhile, the enzyme ceased its activity upon incubation at 45°C beyond 60 min, whereas at 37°C the activity was recorded till 2h. This states that CIAP at alkaline pH from 8 to 11 undergoes slight conformational changes. The enzyme might be attaining kinetically stable conformations at pH 8.5 and 11, as evident from the experimental findings [18].

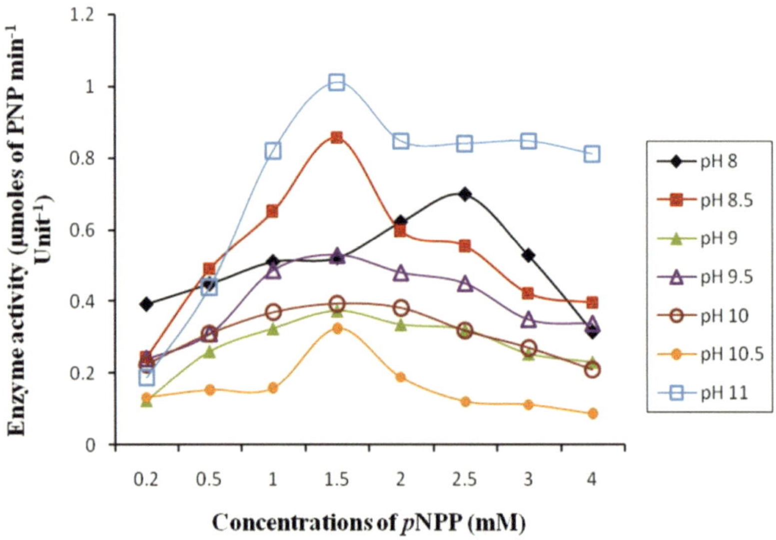

Figure 4. Hydrolysis of pNPP by CIAP under different concentrations of pNPP and varying pH regimes in Tris-HCl buffer.

The effect of substrate concentration on the CIAP activity across a wide range of pH values between 8 and 11 was observed. The observations were in sharp contrast to the observations made by Fernley and Walker [19], who reported inhibition of the enzyme at 0.2–4 mM concentrations of the substrate. From the enzyme-substrate saturation kinetics, it was noted that the saturation of the enzyme for substrate binding occurred at 1.5 mM, regardless of the pH 8.5–11 in Tris-HCl buffer. Inhibition in enzyme activity was observed at increasing substrate concentrations above1.5 mM pNPP at all pH range except pH 11 where stable activity was noticed. At pH 11, dissociation of pNP from pNPP or transfer of the enzyme to another quaternary state might be highly facilitated [18].

3.2. Determination of V_{max}, K_m and K_{cat} at different pH for CIAP activity with pNPP

Over a pH range of 8–11, V_{max} and K_m have been derived from Lineweaver-Burk plot. It was observed that V_{max} and K_m varied along with the pH. The lower the K_m value it indicates the stronger the affinity between enzyme and substrate, whereas the higher the K_m value that reflects the weaker the affinity between them [20]. At pH 11, the V_{max} and K_m of CIAP were determined to be 3.12 μmol min^{-1} $unit^{-1}$ and 7.6×10^{-4} M, respectively. This shows that even the rate of hydrolysis was higher at pH 11, least affinity of the enzyme towards the substrate was observed, which shows the possibility of slight conformational changes in the enzyme [21]. CIAP is reported to have K_m value of 9.6×10^{-4} M for phenyl phosphate [17] while it is found to be 3×10^{-2} M for β-glycerophosphate for rat intestinal-mucosal ALP [22]. By contrast, the K_m values were several orders of magnitude lesser than what is reported earlier and the higher affinity reported here could be partly due to the purified enzyme used [15, 22]. The highest turnover number was found to be 82.98 s^{-1} at pH 11.

3.3. Effect of pH and substrate concentration on BAP activity with pNPP as substrate in Tris-HCl buffer

Buffers play a vital role in the catalytic activity of alkaline phosphatase [23]. In the presence of Tris during the hydrolysis of 4-nitrophenyl phosphate by alkaline phosphatase at pH 8.0, the rate of 4-nitrophenol liberation was increased with concomitant phosphorylation of Tris [24]. On the contrary, it has been discussed that the dephosphorylation of the enzyme would take place at a slower pace under alkaline pH than the catalysis of the substrate. It has also been noted that the activity of BAP, like other enzymes, is pH dependent. BAP upon hydrolysis in Tris-HCl buffer resulted in higher activity at pH 8.5 till pH 10 at 37°C. The activity of enzyme increased linearly with the increase in substrate concentration [18] (**Figure 5**).

3.4. Determination of V_{max}, K_m and K_{cat} at different pH for BAP activity with pNPP

The Michaelis-Menten (K_m) varies from enzyme to enzyme, and also for the same enzyme with different substrates. **Table 1** shows the various kinetic parameters for catalysis of pNPP by BAP in Tris-HCl buffer under different pH. The optimum pH for BAP falls at 8.5 and the V_{max} is 0.82 μmol min^{-1} $unit^{-1}$, and the K_m is 1.5×10^{-3} M. Irrespective of the low rate of hydrolysis, the affinity of the enzyme BAP for pNPP was maximum at pH 8.5. When the K_m for

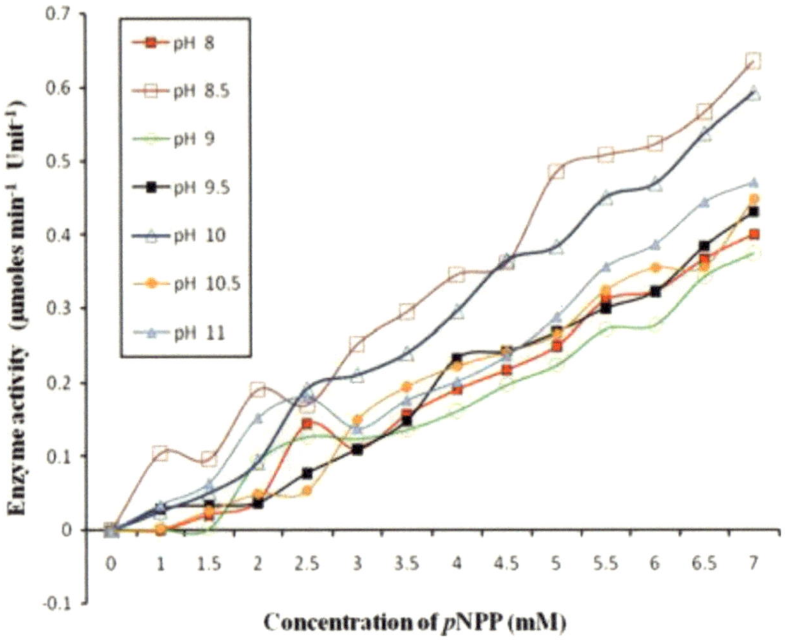

Figure 5. Hydrolysis of pNPP by BAP under different concentrations of substrate and varying pH regimes in Tris-HCl buffer.

pH	V_{max} (µmol min^{-1} unit^{-1})	K_m (M)
8	0.73	7.4×10^{-3}
8.5	0.82	1.5×10^{-3}
9	1.48	3.5×10^{-3}
9.5	0.33	4.66×10^{-3}
10	1.33	1.67×10^{-3}
10.5	0.83	1.75×10^{-3}
11	0.73	7.5×10^{-3}

Table 1. The V_{max} and K_m values obtained with BAP in 50 mM Tris-HCl buffer.

enzyme catalytic reaction is higher, affinity of the enzyme for its substrate is lower and vice versa. Results showed that the catalytic efficiency at pH 8 is 18,220 ± 2 M^{-1} s^{-1} [18].

4. Bioprecipitation studies using calf-intestine alkaline phosphatase

Kinetic parameters obtained for BAP were analysed under standard conditions which showed comparable V_{max} at pH 8, 8.5 and 11, while the values at pH 9 and 10 were alike. The activity of the enzyme was quite complex, and possibly all the parameters during enzymatic precipitation studies with respect to incubation time, temperature, pH and substrate concentration have to be studied [18].

4.1. Average pattern of precipitation of Cr^{6+}, Cr^{3+}, Ni^{2+}, Cd^{2+} and Co^{2+} from single-ion solutions

A comparison of the average graphs for the precipitation of each heavy metal across the two pH regimes 8 and 11 and the concentrations employed, viz., 250 and 1000 ppm, shows that the efficiency of the enzyme-derived reactions is in the order of Cd^{2+}> Ni^{2+}> Cr^{3+}> Co^{2+}> Cr^{6+} (**Figure 6**).

4.2. Bioprecipitation of Cr^{3+} and Cr^{6+} from tannery effluent

The enzyme CIAP was able to precipitate 35.10% of Cr^{6+} and 38.39% of Cr^{3+} from the original concentration of 600 and 350 ppm, respectively, present in the tannery effluent. These precipitations were obtained at 120 min and at pH 11. Further, the percentage of precipitation of Cr^{6+} in these reactions is also comparable to the precipitation observed with Cr^{6+} in single-ion solution earlier (250 ppm, pH 11). This is attributed to the fact that only in the presence of CIAP the precipitation of metal ions could take place.

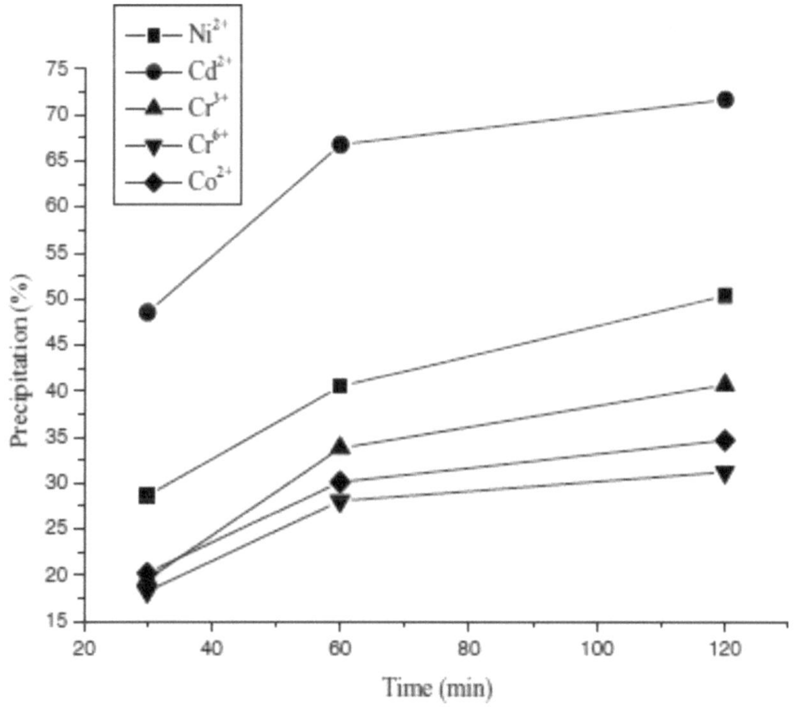

Figure 6. Graphical representation of the average pattern of precipitation of heavy metals from single-ion solution by CIAP and *p*NPP.

4.3. Bioprecipitation of Ni^{2+}, Cd^{2+} and Co^{2+} from electroplating effluent

During experiments with electroplating effluent, it was observed that CIAP was able to precipitate 52.28% of Cd^{2+} from initial concentration of 934 ppm, 31.34% of Co^{2+} from 878 ppm and 30.34% of Ni^{2+} from750 ppm at 120 min at pH 11. The percentage of precipitation of each heavy metal was quite comparable to the observations made with single-ion solutions. However, values obtained were somewhat lesser than what was seen in the former reactions. It could be attributed to the fact that the Pi generated was competed by all the three metal ions simultaneously for the formation of respective metal-phosphate precipitate.

5. Bioprecipitation studies using bacterial alkaline phosphatase

Bacterial alkaline phosphatase precipitated Cr^{6+} and Cr^{3+} from the two different concentrations such as 250 and 1000 ppm at various pH from 8.5 to 10. At 250 ppm, 42 and 52.81% of precipitation for Cr^{6+} and Cr^{3+} were obtained at pH 8.5 during 300 min of incubation. Analysis of the precipitation obtained with BAP under the pH 8.5 and 10, at 250 and 1000 ppm, resulted in differences in the pattern of precipitation. Higher precipitation could be achieved with 250 ppm of Cr^{6+} with 1000 ppm, since the pH seems to play a less important role in determining the precipitation levels. Whereas with Cr^{3+}, the pH influences the precipitation patterns and at both concentrations higher precipitation was obtained at pH 8.5 and 10 [18]. This is because of the dynamics of Cr^{3+} and Cr^{6+} ionization and dissociation as pH dependent rather than the inherent capability of the enzyme in releasing product from pNPP. By determining the statistical data, a significant difference ($p < 0.05$) was observed with a 95% confidence interval, between the two concentrations and pH [18].

5.1. Average pattern of precipitation of Ni^{2+}, Cd^{2+}, Cr^{6+}, Cr^{3+} and Co^{2+} from ion solutions

The average precipitation for every heavy metal between pH 8.5 and 10 at two concentrations 250 and 1000 ppm was compared and is shown in **Figure 7**. The efficiency of the enzyme for the precipitation of different metal ions is in the order of $Cd^{2+}>Ni^{2+}>Cr^{3+}>Cr^{6+}>Co^{2+}$. No scavenging of inorganic phosphate is expected since the precipitations were carried out in ionic solutions, and in addition the amount of inorganic phosphate available for metal phosphate formation should remain constant. Hence, the variation in the precipitation pattern with different ions might be due to the difference in kinetics of metal phosphate complex formation and also because of the radical effect of metal ions present in the backbone of the enzyme [18].

5.2. Enzymatic precipitation of Cr^{6+}, Ni^{2+} and Cd^{2+} from tannery and electroplating industry effluents

The applicability of the enzymatic precipitation of heavy metals from actual industrial effluent was validated by processing the effluent with BAP and pNPP. About 621 ppm of Cr^{6+} was present in tannery effluent. Upon treatment with the enzyme and substrate pNPP for 300 min, the amount of Cr^{6+} present in the supernatant was 403 ppm, resulting in 35.1% of

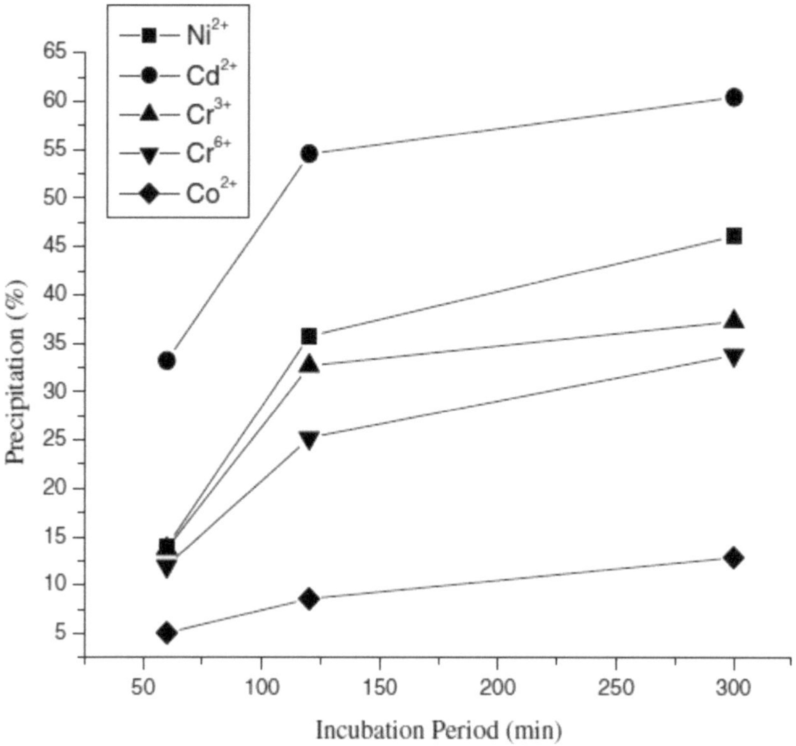

Figure 7. Graphical representation of the average pattern of precipitation of heavy metals from single-ion solutions with BAP using pNPP as substrate.

precipitation. On the other hand, 97 ppm of Ni^{2+} and 122 ppm of Cd^{2+} were also present in the electroplating effluent. Final concentration of 21.43 and 51.95 ppm of Ni^{2+} and Cd^{2+} was present in the supernatant after treating with the enzyme, representing a 77.8 and 57.42% precipitation of the metals, respectively. From the results, it is evident that the two metal ions competed simultaneously for the inorganic phosphate generated by the enzyme. Moreover, Cr^{6+} precipitation from tannery effluent (35.1%) was compared with the values observed in ion solution (42.3%) [18].

The application of BAP and CIAP for the precipitation of heavy metals such as Ni^{2+}, $Cd^{2+,}$ $Cr^{3+/6+}$ and Co^{2+} (from single-ion solutions as well as tannery and electroplating industrial effluents) under alkaline pH was studied using ascorbic acid 2-phosphate, a natural substrate. To determine the potential of the enzyme to remain stable at certain environmental conditions, the kinetic characteristics of BAP and CIAP at various pH 8–11 were also studied. Higher activity of the enzyme was obtained at pH 9.5 and 10 for BAP and CIAP, respectively. The average precipitation pattern of metal ions from single-ion solutions by BAP and CIAP occurred in the order of $Cd^{2+} > Ni^{2+} > Co^{2+} > Cr^{3+} > Cr^{6+}$ and $Co^{2+} > Cd^{2+} > Ni^{2+} > Cr^{6+} > Cr^{3+}$, respectively. Results state that the precipitation of Cr^{6+} from tannery effluent by BAP was 15.57% after 300 min of incubation while CIAP resulted in 71.47% at 120 min incubation. By contrast, BAP precipitated Cd^{2+} at much higher percentage than CIAP from electroplating effluents with the percentage of 94.6 and 66, respectively. Since ascorbic acid 2-phosphate is a natural and biodegradable substrate, this study offers an eco-friendly approach for a sustainable environment [25].

6. Kinetic behaviour of BAP (*E. coli*C90) and CIAP with α-naphthyl phosphate

6.1. Effect of pH and substrate concentration on BAP (*E. coli*C90) and CIAP activity with α-naphthyl phosphate as substrate

The investigation revealed that there is an immense effect of pH as well as the substrate α-naphthyl phosphate concentration on the kinetics of BAP- and CIAP-catalysed α-naphthyl phosphate hydrolysis reaction. The maximum activity of CIAP was found to be at pH 11 (**Figure 8a**) followed by pH 9.5. The second highest pH value after pH 11 was found to be pH 9.5. Experimental data showed that among all pH studied, the highest activity for BAP was found at pH 8.5 (**Figure 8b**). The second highest pH value after pH 8.5 was found to be pH 9. It was observed that with an increase in the substrate concentration, the activity was also being increased. Kinetic parameters such as V_{max} and K_m were calculated for both enzymes. For CIAP, the V_{max} and K_m at pH 11 were 4 μmol min^{-1} unit^{-1} and 2.2×10^{-3} M, respectively. On the other hand, for BAP at pH 8.5 the V_{max} and K_m were found to be 1.25 μmol min^{-1} unit^{-1} and 0.014 M, respectively.

6.2. Average pattern of precipitation of Cr^{6+}, Cr^{3+}, Ni^{2+}, Cd^{2+} and Co^{2+} from single-ion solutions by CIAP and BAP with α-naphthyl phosphate

A comparative analysis of the average graph, obtained on CIAP- and α-naphthyl phosphate-mediated precipitation, was done in order to get a clear picture of the pattern of precipitation. The average precipitation of metals was found to be in the following order: $Co^{2+} > Ni^{2+} > Cd^{2+} > Cr^{3+} > Cr^{6+}$ (**Figure 9a**).

Average graphs for the precipitation of Cr^{3+}, Ni^{2+}, Cd^{2+},Co^{2+} and Cr^{6+} across the two concentrations (250 and 1000 ppm) and the two pH regimes (pH 8.5 and pH 9) are represented in **Figure 9b**. It can be inferred from the comparative analysis that BAP- and α-naphthyl phosphate-mediated heavy-metal precipitation were in the following order: $Ni^{2+} > Co^{2+} > Cr^{6+} > Cr^{3+} > Cd^{2+}$.

6.3. Bioprecipitation of heavy metals from electroplating industrial effluents by CIAP and BAP with α-naphthyl phosphate

CIAP-mediated investigation was performed at pH 11. In electroplating effluent, heavy metals present were Ni^{2+} (initial concentration was 134 ppm), Cd^{2+} (initial concentration was 734 ppm) and Co^{2+} (initial concentration was 278 ppm). The precipitations of Ni^{2+}, Cd^{2+} and Co^{2+} were 66.17, 66.93 and 49.07%, respectively.

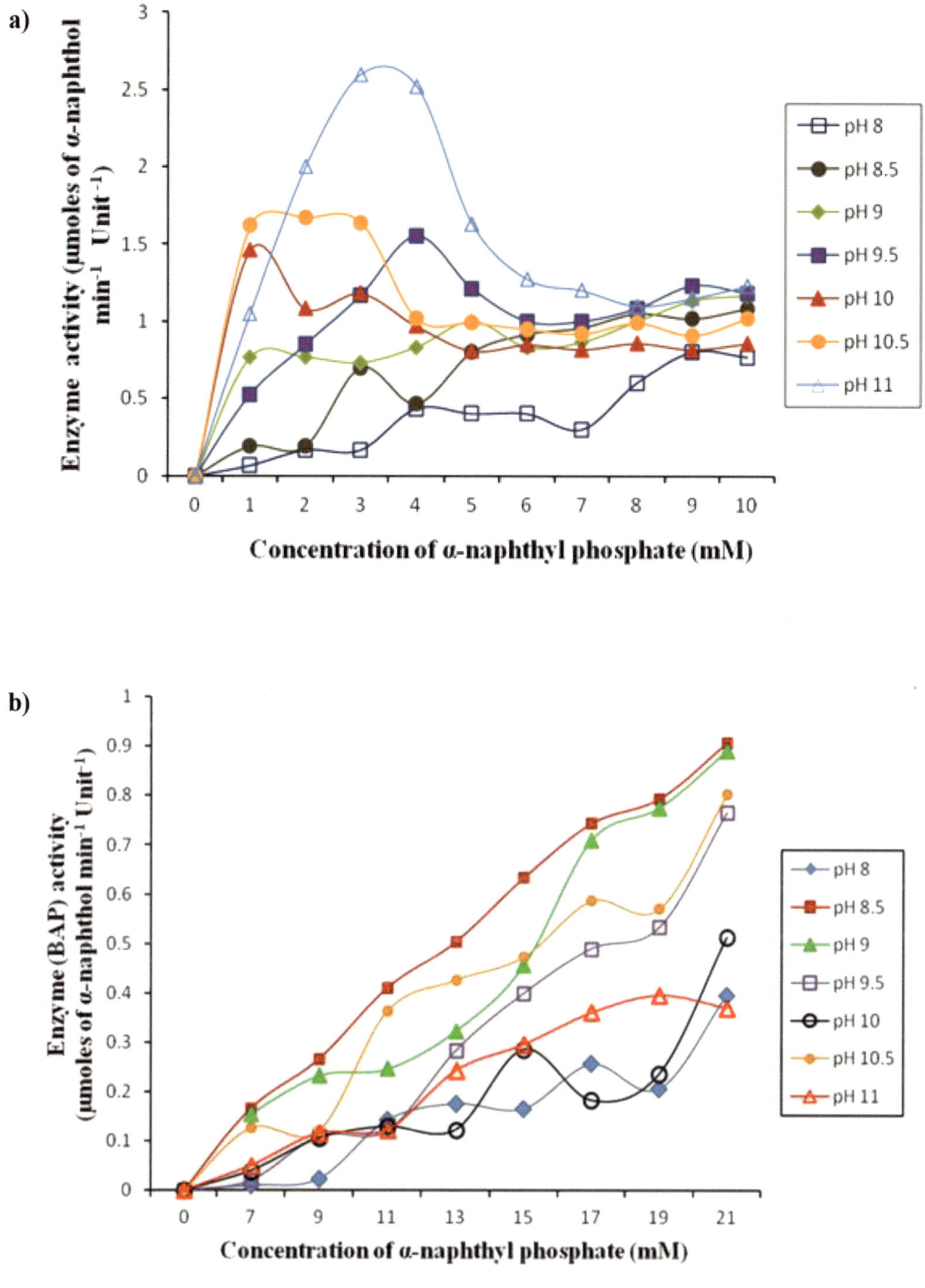

Figure 8. Hydrolysis of α-naphthyl phosphate by (a) CIAP and (b) BIAP under different concentrations of substrates and varying pH regimes.

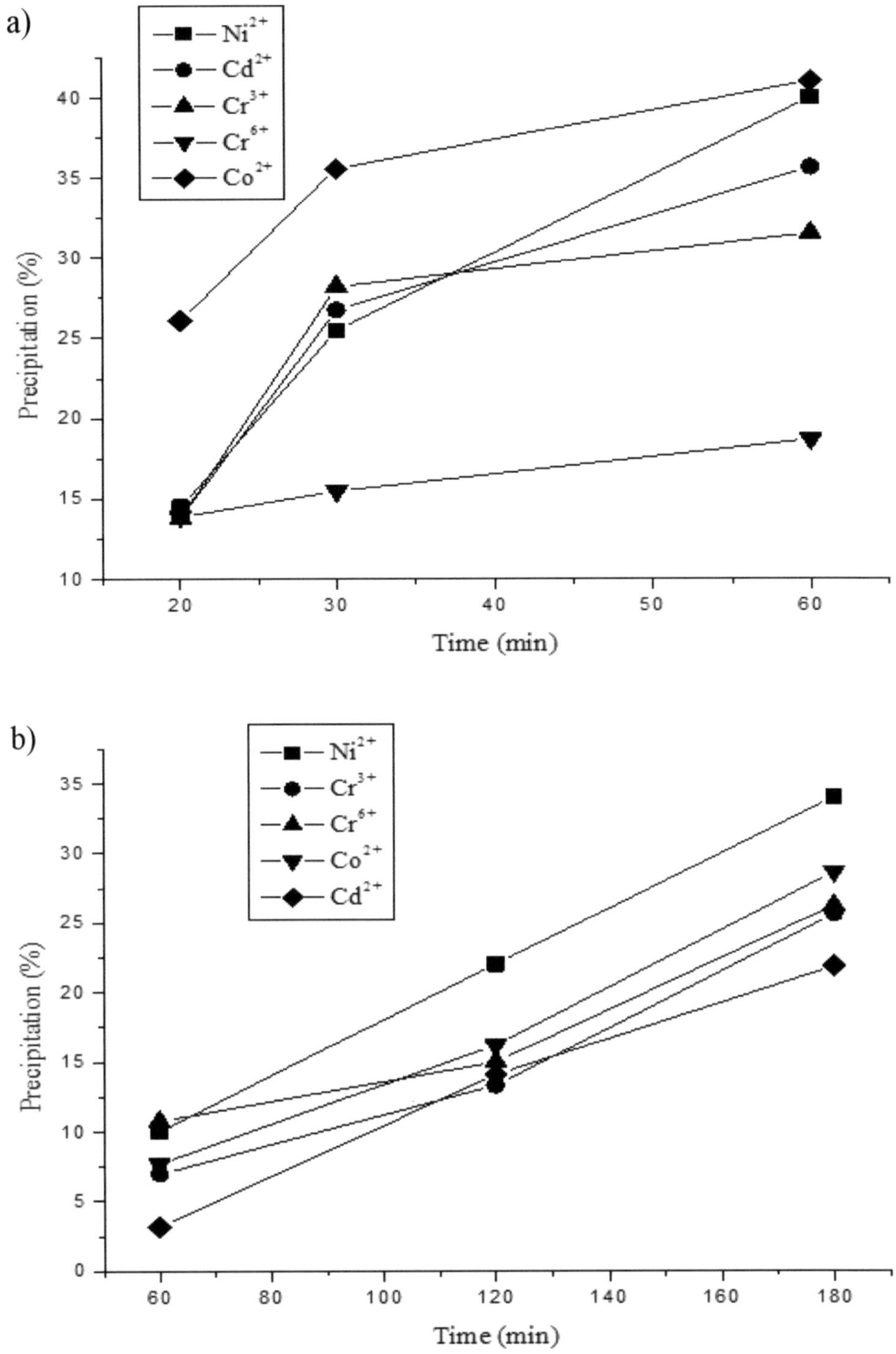

Figure 9. Graphical representation of the average pattern of precipitation of heavy metals from single-ion solution by (a) CIAP and (b) BAP with α-naphthyl phosphate.

On the other hand, the observations showed that BAP-mediated precipitation of Cd^{2+}, Ni^{2+} and Co^{2+} from electroplating effluent was 20.37% (initial concentration was 734 ppm), 57.93% (initial concentration was 134 ppm) and 34.4% (initial concentration was 278 ppm). CIAP with α-naphthyl phosphate-mediated precipitation was greater than that of BAP and α-naphthyl phosphate. This could be attributed to the fact that more P_i had been generated and bound to the heavy metals in the case of CIAP than that of BAP.

7. Kinetic study of BAP (*E. coli*C90) and CIAP with L-ascorbic acid 2-phosphate

7.1. Effect of pH and substrate concentration on BAP (*E. coli*C90) and CIAP activity with L-ascorbic acid 2-phosphate as substrate

The capacity of BAP and CIAP along with ascorbic acid 2-phosphate is being reported as new heavy-metal-remediating tool, the effect of pH and substrate concentration on BAP and CIAP activity was studied at the range of pH values from 8 to 11 and at a range of substrate concentrations from 1 to 17 mM. Generally, tannery and electroplating effluents will have pH of alkaline range. The maximum activity of BAP was observed at pH 9.5 (**Figure 10a**), whereas for CIAP it was at pH 10 (**Figure 10b**).

7.2. Average precipitation of heavy metals from single-ion solution with BAP (*E. coli*C90) and CIAP with ascorbic acid 2-phosphate

A comparative analysis of the average graphs for the precipitation of each heavy metal (Cd^{2+}, Co^{2+}, Ni^{2+}, Cr^{3+} and Cr^{6+}) across the two pH values (pH 9.5 and 10.5 in case of BAP; pH 8 and pH 10 in case of CIAP) and the two concentrations, that is, 250 and 1000 ppm, was employed. The comparative analysis reveals the effect of BAP (*E. coli* C90) on the precipitation of heavy metals which are in the order of $Cd^{2+} > Ni^{2+} > Co^{2+} > Cr^{3+} > Cr^{6+}$ (**Figure 11a**). Similarly in **Figure 11b**, the average graph shows the efficiency of CIAP for heavy-metals precipitation which are in the order of $Co^{2+} > Cd^{2+} > Ni^{2+} > Cr^{6+} > Cr^{3+}$. But at 120 min, the order was different, for example, $Cr^{6+} > Co^{2+} > Cd^{2+} > Ni^{2+} > Cr^{3+}$. The precipitations were carried out in single-ion solutions. Therefore, no scavenging of P_i is expected and the amount of inorganic phosphate liberated for metal-phosphate formation should be constant. In addition, the variation in the precipitation with respect to different ions might be because of the difference in kinetics of metal-phosphate formation and also due to the radical effect of metal ions present in the enzyme [25].

7.3. Bioprecipitation of heavy metals from real-time effluent

The efficiency of BAP and CIAP in precipitating heavy metals from real-time effluents was studied by treating the effluents along with the substrate. The concentration of Cd^{2+} in electroplating effluent initially was found to be 734 ppm. After enzymatic treatment with BAP for 60, 120 and 300 min at pH 9.5, the percentage precipitations obtained were 81.94, 91.39 and 94.62%, respectively, for each time. Likewise, treatment using CIAP for 30, 60 and 120

a)

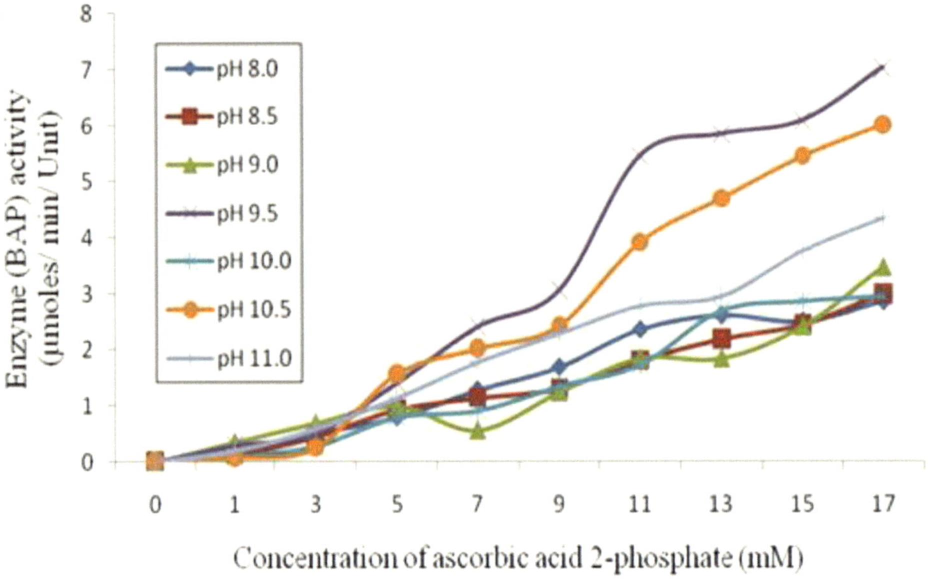

b)

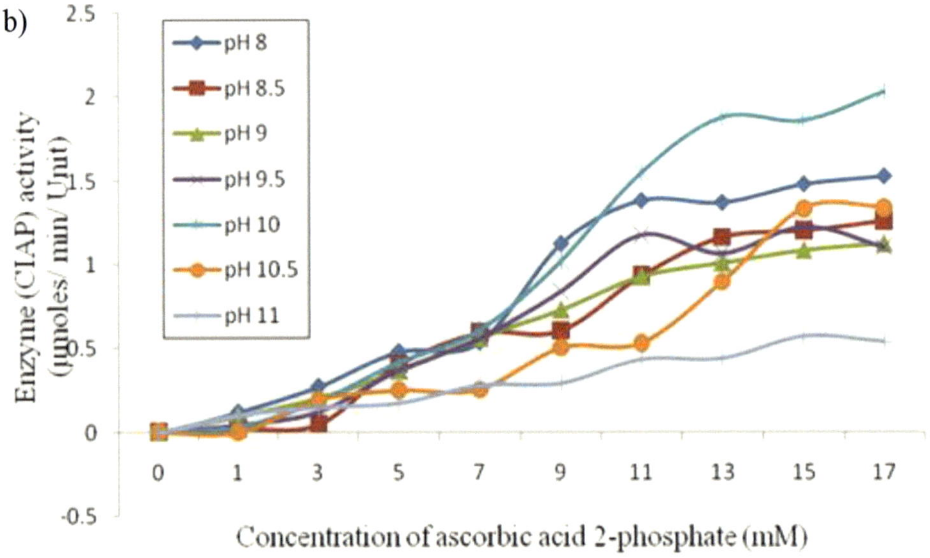

Figure 10. Hydrolysis of L-ascorbic acid 2-phosphate by (a) BAP and (b) CIAP under different concentrations of substrates and varying pH regimes. Reaction conditions: incubation period = 60 min, temperature = 37°C, buffer = 50 mM Tris-HCl.

a)

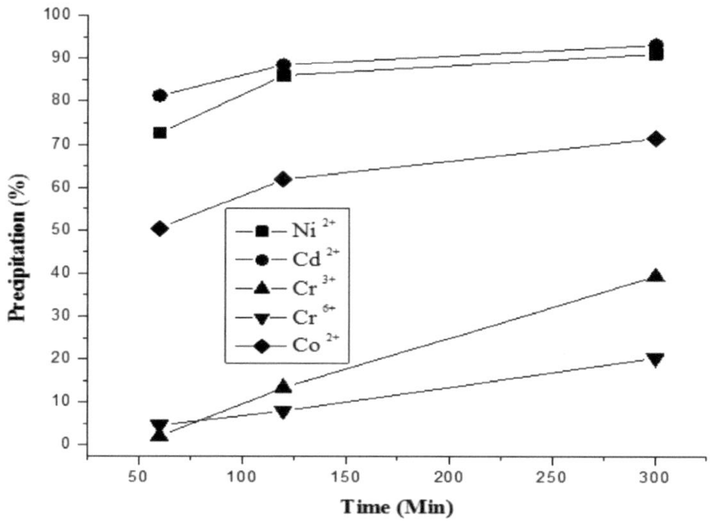

b)

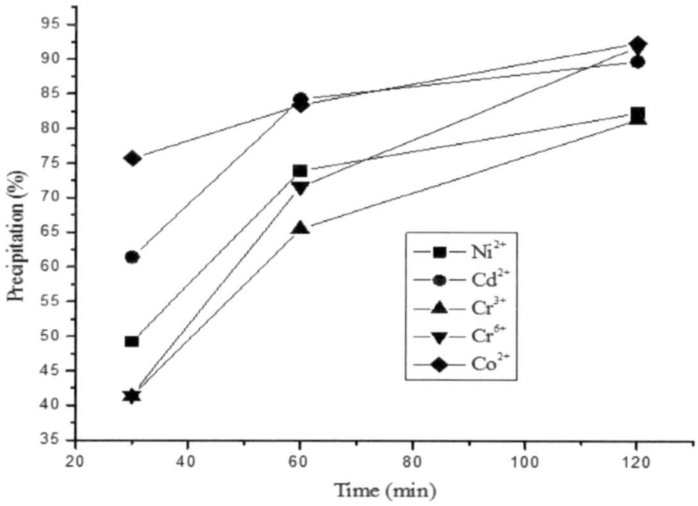

Figure 11. Graphical representation of the average pattern of precipitation of heavy metals from single-ion solution with (a) BAP *(E. coli* C90) and (b) CIAP with ascorbic acid 2-phosphate.

min at pH 10 resulted in 32.75, 53.75 and 65.89% of precipitation, respectively. The removal of Cr^{6+} from tannery effluent at an initial concentration of 560 ppm was analysed by treating with BAP and CIAP using ascorbic acid 2-phosphate as substrate. When compared to CIAP, BAP showed reduced levels of precipitation. Moreover, at pH 9.5 for 60, 120 and 300 min, the percentages of precipitation of Cr^{6+} by BAP were 0, 5.61 and 15.57%, respectively. Whereas at pH 10 for 30, 60 and 120 min, the percentage of removal of Cr^{6+} were 38.32, 64.55 and 71.47%, respectively [25].

8. Conclusion

Bioprecipitation study with BAP and CIAP along with the substrate ascorbic acid 2-phosphate is a much safe and convenient method for the removal of heavy metals such as cobalt, nickel, chromium, cadmium, and so on. The comparative studies have given a clear picture of both the enzymes' kinetic behaviour and bioprecipitation efficiency at different pH, different metal concentration and at different time. Based on this information, further studies can be conducted with other metals which have not been studied in the present work. Immobilization of these enzymes also can be done for large-scale application in industries.

Acknowledgements

The authors are thankful to SRM University, Kattankulathur, Tamil Nadu, India, for providing the facilities to carry out the research work.

Author details

Gouri Chaudhuri[1], Uma Selvaraj[1], P. Venu-Babu[2] and Richard W. Thilagaraj[1]*

*Address all correspondence to: thilagaraj.richard@gmail.com

1 Department of Biotechnology, School of Bioengineering, SRM University, Kattankulathur, Tamil Nadu, India

2 Nuclear Agriculture and Biotechnology Division, Bhabha Atomic Research Centre, Trombay, Mumbai, India

References

[1] Pasqualini S, Panara F, Antoneilli M. Acid phosphatase activity in Pinus pinea Tuber Albidum Ecto Mycorrhizal Association. Canadian Journal of Botany. 1992;**70**: 1377-1383

[2] Beileski RL. Phosphate pool, phosphate transport and phosphate availability. Annual Review of Plant Physiology. 1973;**24**:225-252

[3] Hoylaerts MF, Manes T, Millan LJ. Mammalian alkaline phosphatases are allosteric enzymes. Journal of Biological Chemistry. 1997;**272**:22781-22787

[4] McComb RB, Bowers GN, Posen S. Alkaline Phosphatase. New York, NY: Plenium Press; 1979

[5] Kim EE, Wyckoff HW. Structure of alkaline phosphatases. Clinica Chimica Acta. 1990;**186**:175-187

[6] Kim EE, Wyckoff HW. Reaction mechanism of alkaline phosphatase based on crystal structures: Two metal ion catalysis. Journal of Molecular Biology. 1991;**218**:449-464

[7] Lorenz B, Schroder HC. Mammalian intestinal alkaline phosphatase acts as highly active exopolyphosphatase. Biochimica et Biophysica Acta. 2001;**1547**:254-261

[8] Zhifang C, Zhen X, Yongdoo P, Haimeng Z. Activation of calf intestinal alkaline phosphatase by trifluoroethanol. Tsinghua Science and Technology. 2001;**6**:426-431

[9] Millan JL. Alkaline phosphatases structure, substrate specificity and functional relatedness to other members of a large superfamily of enzymes. Purinergic Signalling. 2006;**2**:335-341

[10] Moss DW. The influence of metal ions on the orthophosphatase and inorganic pyrophosphatase activities of human alkaline phosphatase. Biochemical Journal. 1969;**112**:699-701

[11] Ghosh SS, Bock SC, Rokita SE, Kaiser ET. Modification of the active site of alkaline phosphatase by site directed mutagenesis. Science. 1986;**231**:145-148

[12] Coleman JE. Structure and mechanism of alkaline phosphatase. Annual Review of Biophysics and Biomolecular Structure. 1992;**21**:441-83

[13] Le Du MH, Stigbrand T, Taussig MJ. Crystal structure of alkaline phosphatase from human placenta at 1.8 Å resolution. Implication for substrate specificity. Journal of Biological Chemistry. 2001;**276**:9158-9165

[14] Stec B, Holtz KM, Kantrowitz ER. A revised mechanism for the alkaline phosphatase reaction involving three metal ions. Journal of Molecular Biology. 2000;**299**:1303-1311

[15] Wilson IB, Dyan J, Cyr K. Some properties of alkaline phosphatase from Escherichia coli transphosphorylation. Journal of Biological Chemistry. 1964;**239**:4182-4185

[16] Fernley HN, Walker PG. Phosphorylation of Escherichia coli alkaline phosphatase by substrate. Nature. 1966;**212**:1435-1437

[17] Stinson RA, Chan JRA. Advanced Protein Phosphatases. Vol. 4. Leuwen University Press; 1987. pp. 77-93

[18] Chaudhuri G, Dey P, Dalal D, Venu-Babu P, Thilagaraj WR. A novel approach to precipitation of heavy metals from industrial effluents and single-ion solutions using bacterial alkaline phosphatase. Water Air and Soil Pollution. 2013;**224**:1625

[19] Fernley HN, Walker PG. Kinetic behaviour of calf-intestinal alkaline phosphatase with 4-methylumbelliferyl phosphate. Biochemical Journal. 1965;**97:**95-103

[20] Murray RK, Bender DA, Botham KM, Kennelly PJ, Rodwell VW, Weil PA. Harper's Illustrated Biochemistry. 28th ed. Chapter 8. New York, NY: McGraw Hill; 2009

[21] Chaudhuri G, Chatterjee S, Venu-Babu P, Ramasamy K, Thilagaraj WR. Kinetic behaviour of calf intestinal alkaline phosphatase with pNPP. Indian Journal of Biochemistry & Biophysics. 2013b;**50**:64-71

[22] Ross MH, Ely JO, Archer JG. Alkaline phosphatase activity and pH optima. Journal of Biological Chemistry. 1951;**192**:561-568

[23] Orhanovic S, Pavela-Vrancic M. Alkaline phosphatase activity in seawater: Influence of reaction conditions on the kinetic parameters of ALP. Croatica Chemica Acta. 2000;**73**:819-830

[24] Trentham DR, Gutfreun, H. The kinetics of the reaction of nitrophenyl phosphates with alkaline phosphatase from Escherichia coli. Biochemical Journal. 1968;**106**:455-460

[25] Chaudhuri G, Venu-Babu P, Dalal D, Thilagaraj WR. Application of alkaline phosphatase for heavy metals precipitation using ascorbic acid 2-phosphate as an effective natural substrate. International Journal of Environmental Science and Technology. 2015;**12**:3877-3886.

Chapter 3

Diluted Thermopressurized Phosphoric Acid: A Gentle Proton Donor for Polysaccharide Acid Depolymerization and (Bio)processing

José Domingos Fontana, Marcela Tiboni and
Heidegrid Siebert Koop

Additional information is available at the end of the chapter

http://dx.doi.org/10.5772/intechopen.68236

Abstract

Phosphorus is a very important element for several metabolic pathways in all living organisms as exemplified by DNA, RNA, glucose and fructose-P, and adenosine triphosphate (ATP). The whole metabolism of phosphate in any living organism involves the catalysis carried out by many enzymes, such as kinases, pyrophosphorylases, isomerases, and phosphatases. Symptoms of hypophosphatemia include neurological dysfunction and disruption of muscle and blood cells and could be caused by malnutrition, failure to absorb phosphate, and metabolic syndromes. Phosphoric acid is widely used as an acidifying agent in a variety of pharmaceutical formulations as an acidulant, flavor, and synergistic antioxidant and sequestering. At the laboratorial and industrial territories, due to safety precautions, phosphoric acid may be considered a valid acid alternative for stronger and risky acids such as sulfuric, hydrochloric, and nitric acids. Furthermore, phosphoric acid, among the mineral acids, is less corrosive for steel and all goods made therefrom. Taking into account all these favorable arguments, the applied research at our laboratory (LQBB) is focused, with success, in the utilization of much diluted and moderately thermopressurized phosphoric acid (*o*-PA) in the pretreatment of polysaccharides for many biotechnological, as oligosaccharides production, important prebiotics for the human gastrointestinal tract.

Keywords: phosphorus, phosphoric acid, oligosaccharides, prebiotics, polysaccharides depolymerization

1. Introduction

Henning Brand, a German merchant, when distilling urine discovered phosphorus and historically registered it, in 1669, through a communication to the famous chemist Gottfried

Wilhelm Leibniz. The addition of sand or coal to urine facilitated to turn phosphorus free. However, the first industrial sources for phosphorus production were bones previously converted into phosphate with nitric or sulfuric acid and then again treated with coal. This has been the contribution of Carl Wilhelm Scheele, in 1770. Much before, Arabian chemists had described a similar process also using urine. Etymologically, the word "phosphorus" was taken from Greek ("*phos*" + "*phoros*" = light + bearer) since it can be seen, in the darkness, when elemental phosphorus is exposed to the air [1, 2].

Overpassing several unsuccessful attempts, James Readman, at Edinburgh, Scotland, built an electric furnace where phosphate-rich materials mixed with carbon and silica generated free phosphorus. He transferred the patent rights to Albright & Wilson Co., from Oldbury, England, a company that started the industrial production of phosphorus and then exported the knowledge to France, Germany, Canada, and the United States [3].

Despite many technological improvements and huge scale up, the Readman method still holds and it consists of a strong reducing heating of fluorapatite with coke and sand at 1400–1500°C for the recovery of phosphorus vapors:

$$4Ca_5F\left(PO_4\right)_3 + 30C + 18SiO_2 \rightarrow 18CaO \cdot SiO_2 \cdot 1/9CaF_2 + 30CO + 3P_4 \quad (1)$$

The energy demand of this electrothermic process is very high, around 14 MWh/ ton. Reason by which only a few countries with an abundant supply of hydroelectric power as is the case of Kazakhstan, China, and the United States use to adopt this technology despite the sources of fluorapatite mines being also widespread at other countries such as, Morocco, Tunisia and Russia. Usually, three vertical electrodes feed a modern furnace with the energy high input for the vaporization of P_4 from a mixture of coke, sand, and phosphate rock in a proportion of 16:30:100. A water spray followed by more cold water leads then to the condensation of P_4, so a furnace unit with 12 m diameter and 8 m height can produce around 30,000 ton of white P/year. However, the whole process is not restricted to this short description. The presence of some silica, calcium, and iron in phosphate rock fluorapatite demands some parallel steps to conveniently and environmentally address these minor byproducts like ferrousphosphorus.

The other important P allomorph, red phosphorus, although easily ignited is not spontaneously flammable. Its production from white phosphorus is somewhat simpler: P_4 is maintained under a bed of water inside a closed steel reactor, heated at 288°C for a couple of days. The red product escapes to a lid pipe, followed by a condenser. Since the chemical nature is the same, the representative equation is: P_4 (white) → P_4 (red).

A temperature elevation to 356°C heat allows the distillation of any residual white phosphorus whose boiling point (b.p.) is 289.5°C. Crystalline red phosphorus has a very high melting temperature (590°C).

Incorporation of phosphorus to ignite wooden splints impregnated with sulfur as precursors of the modern matches was carried by Robert Boyle, in 1680. The industrial practice led to a laboral facial deforming disease provoked in the workers by white phosphorus and then named *phossy jaw*.

2. Phosphorus: basic chemical and physical aspects

Phosphorus (P; in the Table of Elements: atomic number = 15; atomic weight = 30.973761998), and a very particular $3s^2$ $3p^3$ shell of electrons, may display as much as eight progressive states of oxidation: +5, +4, +3, +2, +1, −1, −2, −3, and −4 as a direct reflex of its electronic configuration [Ne] $3s^2$ $3p^3$. Phosphorus is so reactive that it allows several binary chemical combinations mostly neglected in the literature. Examples are its combination with halogens such as PBr_3, PCl_5, PI_3, PF_5, with nitrogen PN and P_3N_5 as well other divalent pairs (Ca_3P_2, CuO_2, and FeP/Fe_3P_2). Obviously, the expected combination with oxygen leads to several oxides (P_2O_3, P_2O_4, P_2O_5), in turn generating, upon hydration, the respective oxyacids. Given that sulfur may replace oxygen in some instances, phosphorus also combines with it to P_4S_3 and P_4S_{10}.

Phosphorus (P) is scored as the 12th more abundant element on Earth. In parts per million units, P is around 1050 ppm by weight or 730 ppm by moles of all mineral matter in the soil crust. About 190 different minerals possess P in their composition, the richest one being apatite, a combination with calcium, along with minor proportions of flour, chlorine, or hydroxide in a general formula [$Ca_{10}(PO_4)_6$(F, Cl, or $OH)_2$]. The most important source for industrialization is phosphorite or phosphate rock mainly found in North Arica, Russia, and the United States what account for almost all of the 5×10^{10} tons available in the Earth's crust. P′s single natural isotope is P^{31} although some artificial isotopes through nuclear reactions like P^{32} (the only one stable) are very useful for research purposes given its very short half-time life: only 14.3 days and then the radio mapping of several metabolic pathways in all living organisms as exemplified by DNA, RNA, glucose and fructose-P (the feed for the glycolytic pathway), and adenosine triphosphate (ATP), the energetic coin and P-lipids of any cell membrane. Furthermore, phosphate is the main electrolyte in the human plasma and is between 1.12 and 1.19 mM L^{-1}, and there are around 750 g of P in the human body. P is so reactive that it is not found in the nature as its single element. Rather, when reductively produced in the laboratory, its names are white and red phosphorus. White phosphorus when reacting with air oxygen generates a glow formerly known as *phosphorescence* and nowadays as *chemiluminescence*. Other allotropic forms of the element are "scarlet or violet" and "black" phosphorus, the latter a conductor of electricity. All these P allotropes are crystalline from cubic to orthorhombic types. These are very dense substances with an average d = 1.828–2.69 g cm^{-3}. The tetrahedral white phosphorus given its flammability and pyrophoricity was used as one of the ingredients of "war weapons" that the USA used in Vietnam and other countries under military conflicts.

Phosphorus (P) as arsenic (As) as well as nitrogen (N), bismuth (Bi), and antimony (Sb) is said to be a pnictogen (five electrons in the outermost shell) and its 0.1% natural occurrence make it the most abundant pnictogen among the five in the terrestrial crust [4].

3. Phosphorus-derived oxoacids

The oxoacids derived from phosphorus display three particular atomic aspects: acidic protons and –P-P-links and sometimes, nonacidic hydrogen directly linked to P. The 10 most well-known occurrences or synthetic derivatives as oxyacids from P are:

(a) Valence +5 P (and acidic protons)

- H_3PO_4 orthophosphoric acid (3)
- $H_4P_2O_7$ pyrophosphoric acid (4)
- $H_5P_3O_{10}$ tripolyphosphoric acid (3)
- $H(HPO_3)nOH$ polyphosphoric acid (n + 2)
- $(HPO_3)n$ metaphosphoric acid (n)

b. Valence +4

- $H_4P_2O_6$ hypophosphoric acid (4)

c. Valence +3

- H_3PO_3 (ortho)phosphorous acid (1)
- HPO_2 metaphosphorous acid (1)
- H_2HPO_3 phosphorous acid (1)

d. Valence +1

e. HH_2PO_2 hyphosphorous acid (1)

The most prevalent compounds of phosphorus are derivatives of phosphate (PO_4^{3-}), a tetrahedral anion. Being triprotic, phosphoric acid converts stepwise to three conjugate bases as a consequence of its progressive pK_{a1}, pK_{a2}, and pK_{a3} = 2.15, 7.20, and 12.35:

$$H_3PO_4 + H_2O \rightleftharpoons H_3O^+ + H_2PO_4^- \ldots K_{a1} = 7.1 \times 10^{-3} \quad (2)$$

$$H_2PO_4^- + H_2O \rightleftharpoons H_3O^+ + HPO_4^{2-} \ldots K_{a2} = 6.3 \times 10^{-8} \quad (3)$$

$$HPO_4^{2-} + H_2O \rightleftharpoons H_3O^+ + PO_4^{3-} \ldots K_{a3} = 4.5 \times 10^{-13} \quad (4)$$

4. Phosphorus, phosphoric acid, and its esters: their outstanding importance in the biological life and environment

Although the most intensive if not relevant application of phosphorus is as phosphate-based fertilizers, one should recall that any living matter from a bacterium to the man is vitally dependent on phosphate since a lot of noble and essential biochemical molecules are built on phosphate: nucleic acids (DNA and RNA) as their nucleotide-based backbone is bridged by phosphodiester linkages; the life most important energetic coin, adenosine triphosphate (ATP); glucose- and fructose-P, the feeding fuels for the universal glycolytic, or

Embden-Meyrhoff-Parnas pathway which requires the previous activation by phosphorylation of the neutral hexose pair prior to their entry in EMP pathway; phospholipids in the assembly of cell membranes; and the serine-P differential aminoacid in casein, the perfect food for newborn. As a critical example, cholesterol biosynthesis' (and by extension several hormones derived therefrom) first steps are the sequential building of mevalonate-mono- and di-phosphate. Furthermore, phosphorylation (as well dephosphorylation) of key enzymes through the action of specific, also enzymatic catalysts (kinases), controls the "on-off" of whole metabolic pathways.

Furthermore, most of the metabolic pathways of the human, animal, and microorganism cells and tissues ask for the direct or indirect energy content of ATP (or other coenzyme analogs as NADP—nicotinamide adenine dinucleotide phosphate) to drive particular reaction steps. The energetic content of this coenzyme is elevated:

$$ATP + H_2O \rightarrow ADP + P_i \ \Delta G = -30.5 \text{ kJ mol}^{-1} (-7.3 \text{ kcal mol}^{-1}) \quad (5)$$

$$ATP + H_2O \rightarrow AMP + PP_i \ \Delta G = -45.6 \text{ kJ mol}^{-1} (-10.9 \text{ kcal mol}^{-1}) \quad (6)$$

Human body, on average, contains 250 g of ATP given the permanent coenzyme recycling. Moreover, animal bones and teeth are mainly composed (70%) of a form of hydroxyapatite—$Ca_{10}(PO_4)_6(OH)_2$ but with a Ca:P ratio between 1.37:1.87 instead of 5: 3 as seen in the earth mineral. In a normal adult male human body (80kg), around 12kg are bones (dry weight basis). The proteins of these particular anatomic body pieces experience stiffening thanks to the complexation with calcium phosphate.

The whole metabolism of phosphate in any living organism involves the catalysis carried out by many enzymes. They are designed as kinases (building the phosphoric esters), pyrophosphorylases (synthesizing import sugar donors for polymers), isomerases (changing, for instance, a sugar phosphate hexose, G-6-P, to the similar ketose-6-P), and phosphatases (hydrolyzing phosphate esters). From both the anabolic and catabolic standpoints, the pool of free phosphate, shortly Pi, corresponds to the bioavailability of the cosubstrate for the energetic premobilization of glucose units present in reserve polysaccharides. In this way, glycogen from animals, yeasts, and molds, and starch from plants and algae are sequentially catabolized to glucose-1-P.

Among pathologies of medical interest, hypophosphatemia is a condition of low levels of soluble phosphate levels in the blood serum, and therefore inside cells. Symptoms of hypophosphatemia include neurological dysfunction and disruption of muscle and blood cells due to lack of ATP. Low-phosphate syndromes are caused by malnutrition, by failure to absorb phosphate, and by metabolic syndromes that draw phosphate from the blood (such as refeeding after malnutrition) or pass too much of it into the urine. Conversely, hyperphosphatemia is characterized by too much phosphate leading to diarrhea and calcification (hardening) of organs and soft tissue, and also disturbing the normal body's ability to absorb and utilize iron, calcium, magnesium, and zinc.

Around a half century ago, DIPFP—diisopropylfluorophosphate ($C_6H_{14}FO_3P$.) was in medical use as eye drops do alleviate symptoms of glaucoma-associated ocular hypertension but the consequent eyes opacity precluded this use.

At the molecular level, DIPFP acts as a potent toxin since it combines with serine at the acetylcholinesterase active site thus affecting the normal metabolic interconversion of the neurotransmitter acetylcholine. Furthermore, DIPFP also inactivates some proteases. Its LD_{50} in rat is as low as 6 mg kg^{-1} when administrated orally. This explains its intended use as military weapon by ancient British researchers [5, 6].

Notwithstanding its nomenclature scope, an IUPAC publication gives an excellent sight of the reach of phosphorus/phosphate in natural compounds from any kingdom [7].

5. Natural occurrences of phosphorus and its association with other elements

As a direct consequence from its natural chemical properties and reactivity, the natural occurrence of phosphorus—rocks and minerals—corresponds to the more oxidized forms of the element, the inorganic phosphates (PO_4^{3-} or shortly, Pi), and main fertilizers for plants. Apatite (a calcium phosphate) is its main mineral and does occur with another nature wealth—petrol—this mineral is mainly present at Arabian countries. Florida, USA, Northern Africa, China, and Russia are also important sources of P rocks. The estimate of United States Geological Survey (USGS)—a geological service is around 71 billion tons but at the actual rate of mining and consumption as fertilizer and production of industrial phosphoric acid (1.5 billion ton/year), phosphate minerals may turn exhausted in 4 or 5 decades from 2012. Interestingly, guano (marine seagull and other related birds feces and animal bone ash) have been also explored for phosphorus sources. In fact, the modern technology now thinks about man and animal urine management for the recovery of both phosphate and nitrogen as fertilizers, thus minoring the progressive eutrophication of rivers and lakes. Most of the intensive agriculture relies in the adequate phosphate fertilizers supply to soil. Superphosphate is a special formula for such a purpose and it corresponds to a blend of calcium dihydrogen phosphate [$Ca(H_2PO_4)_2$] and calcium sulfate dehydrate ($CaSO_4{\cdot}2H_2O$). P^{31} is a natural stable isotopic occurrence of phosphorus that is radioactive and is very useful for the investigation of phosphorylated compounds through the spectroscopic technique known as ^{31}P-NMR (nuclear magnetic resonance). Conversely, radiochemistry takes advantages of radioactive occurrences of phosphorus as β-emitters. For instance, P^{32} and P^{33} with half-lives of 14.3 and 25.4 days and energies of 1.71 and 0.25 MeV is useful for research on DNA and RNA probes and their respective Northern and Southern blots as well as DNA sequencing.

6. Production and direct uses of phosphorus

Among other allotropic forms of phosphorus, only two forms have commercial significance—white and red phosphorus—the former even more important given it accounts for around

99% of the world demand. Some food and pharmaceutical processing as well for etching semiconductors require high purity phosphoric acid (H_3PO_4; more strictly designed as *ortho*-phosphoric acid, usually at 85% w/w) also known as "thermal phosphoric acid" which is made by burning phosphorus in moist air. Conversely, phosphoric acid destined to the huger demand of fertilizer production does not require high purity. Hence, it is directly obtained from rock phosphate ores without the intermediary step of elemental P.

Historically (latter half of nineteenth century), the alternative source of phosphate was guano (the excrement of seabirds and bats accumulating in isolated islands as in Chile), a major form of P-based fertilizers lime superphosphate, a mixture of two salts, calcium dihydrogen phosphate $Ca(H_2PO_4)_2$ and calcium sulfate di-hydrate $CaSO_4{\cdot}2H_2O$, produced by the reaction of sulfuric acid and water with calcium phosphate.

The profile of the participation of phosphorus derivatives/products in the industrial activity is shown in **Figure 1**.

Some forms of phosphorus, although low in demand, have very specific applications. For instance, red phosphorus is employed in pyrotechnics and matches and even as fire retardant in plastics like polyamides as it quickly consumes oxygen, the natural oxidizer in burning. World annual production of elemental phosphorus accounted (in 2013) for 850,000 tons [8].

White phosphorus is used as a deoxidizing agent in the preparation of steel and phosphor bronze. Phosphoric acid is primarily used in the production of phosphate compounds. It is also used in pickling metals, in sugar refining, and in soft drinks. Phosphorus forms a number

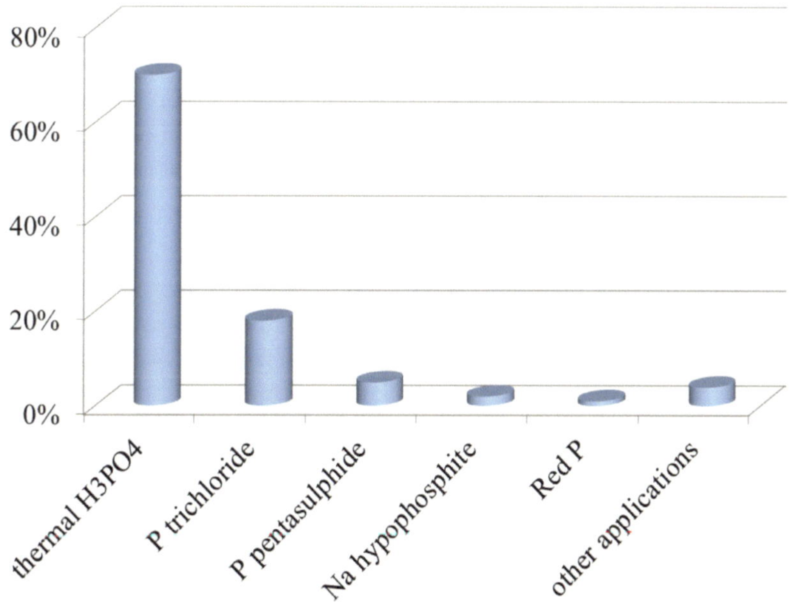

Figure 1. Industrial uses of phosphorus and P-derivatives.

of compounds with halogens, e.g., the trichloride, PCl_3, and the pentachloride, PCl_5, both used as reagents. It also forms an oxychloride, $POCl_3$. It reacts with sulfur to form a pentasulfide, P_2S_5, and a thiochloride, $PSCl_3$, used in insecticides and oil additives. Phosphine, PH_3, is a very poisonous gas.

The lighting effects and colors of fireworks came from their chemical components under burning. Aluminum generates white and silver flames; antimony intensifies bright color, while calcium and barium lead to orange and green colors. Magnesium, copper, lithium, and sodium create bright white, blue, red, and yellow colors, respectively. Iron, depending of the heating intensity, converts into brown to red brown sparks. Zinc is the component for generating smoke. Powdered carbon, along with some sulfur, is the fuel for fireworks and the energy of its burning acts as propellant for all other colors and forms generating components. Herein, phosphorus may turn a key element giving to the night darkness a special glowing effect and also helping the propellancy.

In the dentistry field, trimetaphosphate finds use to reduce the fluoride content of dentifrices without losing its anticariogenic properties. Mouth environmental pH, if reduced, increases the deposition of both fluoride and phosphate on teeth enamel. A reduced mineral loss from teeth results from the incorporation of sodium metaphosphate in dentifrices. The role of several phosphate salts alone or in combination with fluoride has been extensively reviewed and the benefits stand very well established [9–12].

Phosphorus-derived compounds have a wider range of industrial applications. One example is triaryl phosphates in the improved properties they confer to hydraulic fluids concerning lubrication and fire retardation [13].

High-pressure and antiwear properties in EP additives (extreme pressure) in greases, gear oils, and motor oils are benefits arising from oil-soluble organophosphates, with or without zinc, since they provide corrosion protection especially in presence of chlorinated hydrocarbons. For instance, the coordination compounds zinc dialkyldithiophosphates (ZDDP), also antioxidants, start decomposing at 130–170°C, while the activation temperature of tricresyl phosphate (TCP) typically exceeds 200°C. Their reaction products form a chemically bonded lubricating film on the surfaces. TCP has other uses as well: plasticizer in nitrocellulose, lacqueurs, and varnishes and flame retardant in plastics and rubbers [14].

Due to huge material losses (and human, as well) in case of domestic, industrial, and field fires, phosphorus-based flame retardants are being actively developed to provide more appropriate substitutes for the classic brominated flame retardants. So, aryl phosphate dimers and oligomers turned appropriate for plastic flames such as polycarbonate- and polyphenylene-styrene blends. Polyurethane flames are better controlled with an oligomeric aliphatic phosphate. Moreover, polymeric phenylene methylphosphonate and cyclic phosphinates fit better from fires coming from epoxy resin-impregnated boards [15].

Concerning the extraction/purification technologies for crude phosphoric acid, many of them are available for the solvent-based approaches and they are known as the name of their proponents/inventors/factories: Albright & Wilson; Budenheim, FMC, IMI, Prayon, Rhone-Poulenc, Bateman-Wengfu/Prado-AFB Turkey [16].

6.1. Applications and uses of aluminum phosphate A: $Al(H_2PO_4)_3$

Aluminum is one of the most common metals found in nature in the form of the minerals based on aluminum phosphate (e.g., angelate, coeruleolactote, evansite, lucinite, sterretiote, variscites, wavellite). Aluminum phosphate is formed when the phosphoric acid reacts with alumina present in the catalyst. Its industrial applications are varied: as cement in admixture with calcium sulfate and sodium silicate, as a flux for ceramics, in dental cements, waterproofing concrete, as a flame retardant, as a catalyst in organic synthesis, and for special glasses. Aluminum phosphate is also employed as emollient in cosmetics, reliever stomach ailments in the case of heartburn or peptic ulcer pain, and to control excess sweating. Environmentally, the effects of aluminum phosphate are not classified as an acute or chronic toxin to aquatic life or the environment. As an inorganic substance, aluminum phosphate is not biodegradable. Neither bioaccumulation nor biomagnification are considered to be environmental issues [17].

7. The massive agricultural use of a special form of phosphorus: phosphonate-based herbicides

Although a subject that evokes permanent controversy, one of the most intensive industrial applications of phosphoric acid is the chemical synthesis of glyphosate or N-phosphonomethylglycine, invented, produced, and distributed throughout the world by Monsanto with the commercial name "RoundUp®" [18].

It is an interesting and selective herbicide against large-leaf and grassy weeds. In tillages based in the transgenic soya "RoundUp Ready," which are not sensitive to glyphosate, are thus easily turned free from its common weeds thanks to the nebulization with glyphosate. From the chemical standpoint, in glyphosate it occurs as one of the strongest chemical links—phosphonate—a direct linkage between C and P. Since glycine composes most of the proteins and enzymes and has other important anabolic roles in any live organisms (one C donor, precursor of glutathione, creatine, ham, and purine nucleosides), it is comprehensive how its chemical modification toward a phosphonate derivative is so efficient herbicide [19].

Historically, and at the lab scale, glyphosate was synthesized as follows:

$$PCl_3 + H_2CO \rightarrow Cl_2P(=O){-}CH_2Cl \tag{7}$$

$$Cl_2P(=O){-}CH_2Cl + 2\,H_2O \rightarrow (HO)_2P(=O){-}CH_2Cl + 2HCl \tag{8}$$

$$(HO)_2P(=O){-}CH_2Cl + H_2N{-}CH_2{-}COOH \rightarrow (HO)_2P(=O){-}CH_2{-}NH{-}CH_2{-}COOH + HCl \tag{9}$$

Nowadays, glyphosate's industrial production is attained by one of the following pathways, shown in **Figure 2**.

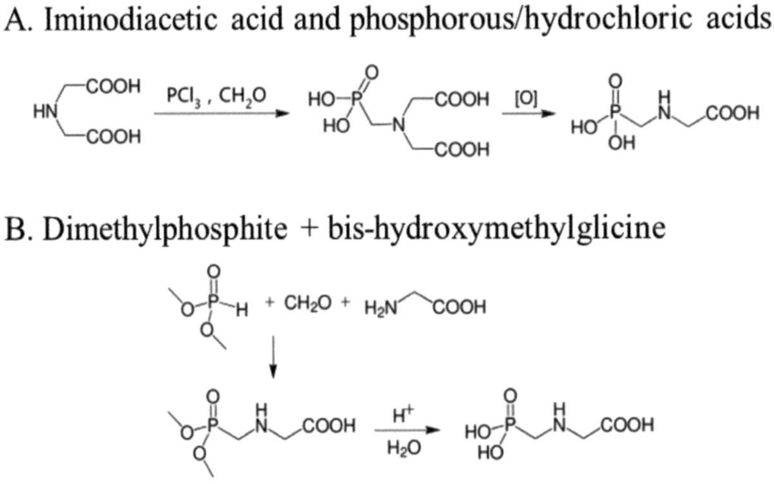

Figure 2. Modern routes for industrial production of glyphosate (N-phosphono-methyl-glycine).

Some of the aspects previously mentioned as controversy can be drawn from three recent review. A 2013 review found that neither glyphosate nor typical glyphosate-based formulations (GBFs) pose a genotoxicity risk in humans under normal conditions of human or environmental exposures. A 2014 review article reported a significant association between B-cell lymphoma and glyphosate occupational exposure. In March 2015, the World Health Organization's International Agency for Research on Cancer classified glyphosate as "probably carcinogenic in humans" (category 2A) based on epidemiological studies, animal studies, and *in vitro* studies [20–22].

There is a plenty of other P-containing herbicides and insecticides—generally said, pesticides sharing the double Janus-faces. On one side, ensuring better crops and agriculture productivity; on the other, scavenging the health of people managing with or consuming contaminated crops, animal food, and water. Other organophosphates (OPI) maintain as the most widely used group of insecticides in the world, despite their toxicity not restricted only to target harmful insects in the agribusiness. Human and animal are the subsequent victims due to the accumulation in soil and waters and their further feeding of contaminated vegetables, fruits, milk, food products, and other living organisms, and it is not an important concern as the profits are coming from the agricultural activity. OPI's main target is the inhibition of the enzyme acetylcholinesterase (ACE), a key enzyme in the ana- and catabolism of the most important neurotransmitter, acetylcholine (Ach), in the brain/nervous system of animals. Following the accumulation of Ach, an overactivation of cholinergic receptors does occur at the neuromuscular junctions as well in autonomic and central nervous system [23].

The general chemical structure of OPI is as follows (**Figure 3**):

Malathion and Dichlorvos, encompassing chemical combination with sulfur and chlorine, respectively, and two examples of most utilized organophosphate pesticides. About a decade ago, the agricultural market used to offer >900 pesticides and >600 active pesticide ingredients.

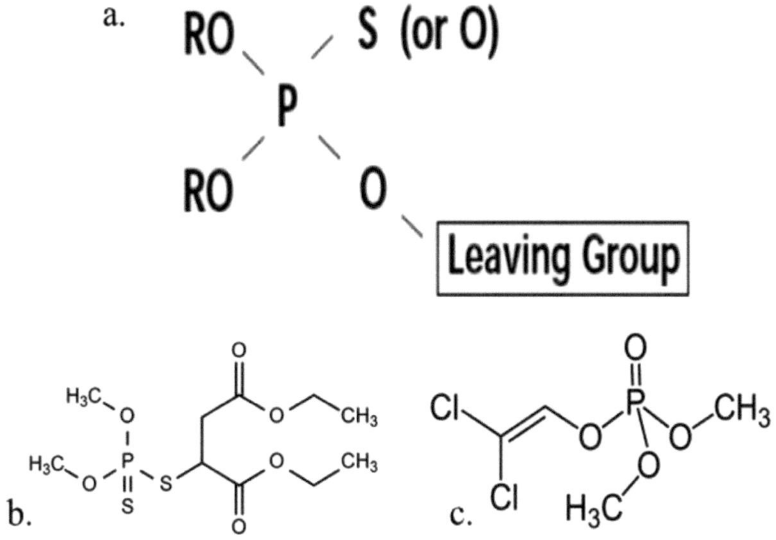

Figure 3. Chemical structure for organophosphorus insecticides (OPI).

Acute toxicity of OPI varies widely. Following is the example of increasing intoxication, and expressed as LD_{50} ranges, chlorfenvinphos (1–30) > dichlorvos (30–50) > malathion (60–1300 mg/kg body weight) [24].

Another particular utilization of phosphorus (white P) is for the production of military weapons such as grenades, mortar shells, and artillery shells for related purposes such as smoke screens for troops movement, marking targets, bullets path tracing, and, worse, as incendiary ammunitions. Viet Cong Tay Ninh province in 1967 and Fallujah at Iraq (2007) experienced the destroying power of these U.S. ancillary weapons. It is worth mentioning that mixtures of phosphoric acid with nitromethane are explosive [25].

8. Environmental utilization of phosphorylated matrices

Phosphorylation of several matrices for multiple purposes as environmental remediation of cationic residual compounds (e.g., cationic dyes and antibiotics) may proceed in the cases of acid-resistant materials (e.g., activated carbon) through the treatment of previously oxidized matrix with a mixture of sulfuric and phosphoric acids or even with the latter alone, provided it is converted in polyphosphoric acid through the chemical removal of the usual moisture of H_3PO_4 85% with the stoichiometric amount of the P anhydride, P_2O_5. For more delicate matrices (e.g., cellulose), a gentle alternative is the utilization, for the same purpose, o cTP—cyclic-Tri-Phosphate, which is easily prepared from potassium dihydrogen phosphate in a muffle at 450°C for 6 h. The next phosphorylating step is carried out at room temperature for 1 h at a moderate alkaline solution with pH 9–10 leading so to the desired products such as phosphorylated activated carbon or cellulose (AC-Pi^{2-}, cellulose-Pi^{2-}), and so on. Polyphosphoric acid is useful for as much as eight different industrial applications: cyclization reactions, rearrangements, dehydration, hydrolysis, polymerization, solid phosphoric acid catalyst, polyamide yarns processing, and polymerization [16].

Particular applications of phosphoric acid in activation in carbonaceous matrices may provide interesting solutions for some contaminations, as is the case of water by the ammonia from human and animal urine. So, activated charcoal impregnated with H_3PO_4 is efficient to remove and modify ammonia gas and hydroxide according to Ref. [26]:

$$H_3PO_4 + 3\,NH_3 \rightarrow (NH_4)_3PO_4 \tag{10}$$

In fact, besides sulfuric acid, phosphoric acid is the common chemical tool to convert impure mined charcoal into activated charcoal [27, 28]. In an opposite sense, excess of phosphate salts discharged in aquatic streams and lakes leads to eutrophication and sometimes to microalgal blooms, as a result of the rupture of normal and harmonic life cycles of several microorganisms [29].

9. Biochemical and pharmaceutical uses of phosphoric products

Concerning the utility of phosphate salts in more sophisticated biochemical applications, once the ionic liquid choline dihydrogen phosphate was demonstrated as an improver of thermostability and shelf-life of several proteins and hence its usefulness as an excipient for the formulation of protein-based pharmaceuticals, the toxicity of several analogs were assayed using a murine macrophage cell line. EC_{50} values were found to be closely related to safe salts (e.g., choline chloride), which indicated feasibility for the intended therapeutic proposal provided the anionic counterpart is not too much larger [30].

Phosphoric acid is also widely used as an acidifying agent in a variety of pharmaceutical formulations. It is used in pharmaceutical products as part of a buffer system when combined with a phosphate salt such as sodium phosphate, monobasic, or dibasic. It is also widely used in food preparations as an acidulant, flavor, and synergistic antioxidant (0.001–0.005%) and sequestering.

Therapeutically, dilute phosphoric acid has been used well diluted in preparations used in the treatment of nausea and sickness. Phosphoric acid 35% gel has also been used to etch tooth enamel and to enhance delivery of drugs through the nail. Nanosized hydroxyapatite powder was made by combining phosphoric acid with eggshells.

The lowest lethal oral dose of concentrated phosphoric acid in humans is reported to be 1286 mL kg^{-1}. LD_{50} (rabbit, skin): 2.74 g kg^{-1} and LD_{50} (rat, oral): 1.53 g kg^{-1}.

In the UK, a 1 in 330 aqueous solution of phosphoric acid is approved as a disinfectant for foot-and-mouth disease. A specification for phosphoric acid is contained in the food chemicals codex (FCC). The EINECS number for phosphoric acid is 231-633-2. The PubChem Compound ID (CID) for phosphoric acid is 1004 [31].

Tableting is one of the largest applications of calcium phosphates in the pharmaceutical field. Due to the different surface areas, calcium slats of phosphoric acid strongly differ concerning their physical properties. For instance, at 93% relative humidity, most of hydroxyapatites can absorb 15% of moisture while the nonhigroscopic basic calcium phosphate dehydrate absorbs <1% [32].

A net mineral loss in the teeth led to porosity, white-spot lesions, caries lesions, and cavitation. The modern approach to circumvent these inconveniences is the remineralization through the application of nanoscale-structured calcium phosphate prepared via the spray-drying technology and the results are encouraging. A typical formulation is calcium and phosphate concentrations around 8 and 5.333 mM L^{-1} after the appropriated blending of anhydrous dicalcium phosphate and calcium carbonate in diluted acetic acid [33].

Given its three strategic dissociation constants (pK_{a1} = 2.15, pK_{a2} = 6.82, and pK_{a3} = 12.38), phosphoric acid ensures one of the widest ranges of buffering capacity for laboratory buffers and biological media through the appropriated mixture of its Na or K salts. Examples are McIlvaine (citric acid and Na_2HPO_4 from pH 2.6 till 7.6), Gomory/Sorensen (Na_2HPO_4 and NaH_2PO_4 from pH 5.8 till 8.0), Clark and Lubs (KH_2PO_4 and NaOH from pH 5.80 till 8.0), Bates and Bower (Na_2HPO_4 and NaOH from pH 11.0 till 11.90) [34].

9.1. Bisphosphonates and its odontological use

The sodium salt of alendronic acid is nowadays a bisphosphonate commercial product utilized to correct several bone diseases such as osteoporogenesis and osteogenesis imperfecta. Its beneficial effect arises from the inhibition of osteoclast-mediated bone resorption without effect on bone remineralization as is the case of pyrophosphate. The only aid is to avoid a parallel hypocalcemia [35–37].

9.2. Uses of phosphorus nonoxygenated compounds

The main utilization of phosphoric acid is the manufacture of fertilizers. Mono- and di-acedi calcium phosphates are employed as additives in the fabrication of toothpaste, animal food, baking powder, and fertilizers. $Na_5P_3O_{10}$ is preferentially used in detergents. Conversely, phosphorus trichloride (PCl_3) and its oxygenated derivative $POCl_3$ are used in the manufacture of pesticides and plasticizers. For the former proposal, phosphorus sulfide (P_4S_{10}) is used as well.

10. Phosphoric acid and phosphate salts in foods and beverages

Phosphoric acid (as a mild acidulant) and phosphate salts are also used to control pH of processed foods. They are also used in medicines for constipation and to prepare the bowel for medical procedures. Two of the most remarkable commercial products where flavor intensity (e.g., arising from sweeteners) is valorized by phosphoric acid or its mix with phosphate salts are the colas and fermented milk, as shown in **Figure 4**.

Phosphoric acid | Tricalcium phosphate

Figure 4. Illustrations for beverages (colas) and fermented milk (yogurt) and their phosphoric acid or P-salts or citric acid ingredients.

In the bakery business, sodium phosphates are often used as emulsifiers, thickening agents, and leavening agents for baked goods. They are also used to control pH of processed food procedures.

Syrup phosphoric acid is very useful to stop or correct oxidized metallic pieces made of iron or steel. Acid application leads to the formation of a barrier pellicle of iron phosphate and hence its antirust action.

Trisodium phosphate is used as a cleaner, water softener, and scale/corrosion inhibitor. Bone ash (calcium phosphate) is used to make chinaware and to make monocalcium phosphate for baking powder.

Sodium tripolyphosphate, shown in **Figure 5**, is used in laundry detergents in some countries, but banned in other countries. It is useful for softening water to enhance the performance of the detergents and to prevent pipe/boiler tube corrosion.

The use of phosphate salts which nowadays turned completely obsolete is in the fabrication of sodium lamps since their replacement by fluorescent lamps and light-emitting diodes (LED) is now consolidated for the sake of energy expenditure cutting.

11. Utilization of diluted, heated, or thermopresurized (ortho) phosphoric acid (*o*-PA) as a mild catalyst for an advantageous phytobiomass polysaccharide partial or total depolymerization

At the laboratorial and industrial territories, due to safety precautions, phosphoric acid may be considered a valid acid alternative for stronger and risky acids such as sulfuric, hydrochloric, and nitric acids, since its more common commercial form, 85% (w/w) H_3PO_4 is really safer.

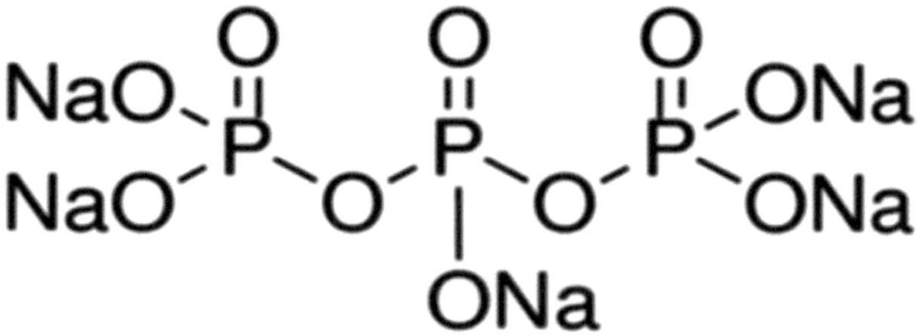

Figure 5. Sodium tripolyphosphate formula.

HCl (37% w/w) and HNO_3 (65% w/w) both evolve as very harmful vapors to the eyes, skin, and lungs while H_2SO_4 (96–98% w/w), given its deep avidity for water, leads to quick carbonization of organic matter and severe wounds to the body, in case of accidents. Another very common utilization of phosphoric acid is in the formulation of buffers (mixes of its Na^+ or K^+ mono- and di-acid salts) as it is equally true for its buffering occurrence in the blood stream of any animal species. Furthermore, phosphoric acid, among the mineral acids, is less corrosive for steel and all goods made therefrom.

Taking into account all these favorable arguments, the applied research at our laboratory (LQBB—Biomasses Chemo/Biotechnological Laboratory at UTFPR—Federal Technological University of Paraná, Curitiba-PR, Brazil) is focused, by decades, in the utilization of very diluted and moderately thermopressurized phosphoric acid (*o*-PA) in the pretreatment of polysaccharides for many biotechnological purposes. Sequentially, the selected substrates were and are hemicelluloses (hetero-xylans from cane bagasse), inulin (a labile polyfructose from dahlia tubercles), cassava starch, and, more recently, hetero-mannans from conifer woods sawdust and commercial seed gums. Homo- and heteropolysaccharides present in phytobiomasses display different lability when facing hot concentrated or diluted mineral acids as consequence of the carbohydrate units (hexoses, pentoses, deoxysugars, acid, and animated hexoses) and even more from the type of glycosidic links between monosaccharide units and their anomericity. For instance, the β-1,4 links of the pyranosidic links of glucose units in cellulose and N-Acetil-glucosamine, respectively, render both very resistant to the hydrolytic action of acids. Usually, a prestep of the polysaccharidic substrate in cold concentrated acid (e.g., 60% w/w), followed by acid dilution (2–4 M) and heating allows the best free monosaccharide recovery. Obviously, the mildest alternative is the use of specific enzymes, namely, cellulases and chitinases, respectively.

In the following descriptions, the authors' experience on the monomerization of polysaccharides contemplating the extremes of natural examples of acid-lability: inulin (the extremaly labile poly β-2,6-fructofuranosyl); starch (a moderately labile poly-α-1,4-gucopyrasyl with a few α-1,6-branches); heteroxylan (also moderately resistant β-1,4-poly-D-xylopyranosyl backbone) with a few of single α-1,3-L-arabinofuranosyl extremely labile lateral units, the same for a few very resistant (4-O-methyl)-D-glucuronopyrasyl units, thus generating aldobiuronic acid); and cellulose (the outstanding resistant and linear polymer of β-1,4-linked glucopyranosyl units) is shown. For any of these particular cases, the diluted *o*-PA acid solutions are better expressed in terms of their effective pH before hydrolysis (after a certain time of equilibration with phytobiomass components) instead of the usual concentration parameters as % (v/w, w/w or mM L^{-1}). As a clever laboratory practice, the explored kinetical and severity parameters may optimize varying phytobiomass mass: diluted *o*-PA (e.g., till 40% in case of inulin and starch; only 10% in case of native ligno(hemi)cellulosics), time of residence in the reactor and more importantly the above-mentioned effective hydrolysis pH (e.g., 3.5–1.5) combined with the thermopressurization (peak temperature and corresponding pressure expressed in atmospheres) vary variable in the range from 60 to 200°C.

11.1. Inulin

Inulin, usually from *Dahlia* tubercles (although also present in chicory, yacon potatoes, and Jerusalem-artichoke roots), with an average degree of polymerization (DP) = 33, is not soluble in cold water as opposed to its complete solubilization in warm water. For pure inulin preparation, tubers from *Dahlia* spp. (e.g., *pinnata*) are washed to remove adherent soil particles, peeled, and quickly submerged in boiling water previously buffered with 25 mM pH 7.0 sodium phosphate to avoid any damage in the polyfructose native structure and the browning from phenoloxidases action. The hot suspension is comminuted in a Waring blender and the suspension is filtered through a 4× layers of cheese cloth. Upon refrigeration at 4°C overnight, inulin settles and may be recovered by centrifugation and dehydration with absolute ethanol and then acetone. The pure polysaccharide should be a white powder or at most, light cream. If still with a more brown hue, its warm solution may be filtered by a bed of $DEAE^{+}$-cellulose for the removal of any residual colored matter.

Inulin is unique among all natural polysaccharides since it lacks a reducing end given its biosynthesis starts from a single sucrose unit (**Figure 6**). In fact, the presence of this minute amount of free glucose in any inulin hydrolysate is the trustable indicative one is managing the intact polysaccharide structure, as depicted. This may be easily confirmed by ^{13}C-NMR since the smaller signals of the single glucopyranose units can be surely detected among the major signals arising from the fructofuranose main units as shown in **Figure 7**.

Inulin partial acid or enzymatic hydrolysis may be quickly monitored even for several samples by thin layer chromatography (TLC) as shown in **Figure 8**. A sharper profile of Fructooligosaccharides (FOS) generated upon *o*-PA or citric acids moderated and partial hydrolyses at pH 2.5 and 85°C for 5 or 15 min is gained with a high performance liquid chromatography (HPLC) shown in **Figure 9**.

Our factorial design prospection indicated the ideal conditions of inulin with *o*-PA regarding the obtention of the higher ratio FOS [FOS: free fructose + hydroxymethylfurfural (HMF)]: pH 2.5–2.0, temperature from 85 to 95°C (preferably the former), and shorter residence times of reaction from 15 to 25 min. Interestingly, an alternative acid catalyst, citric acid, could afford similar hydrolytic results, provided small enhancements of the hydrolysis severity parameters were applied. It will be noted that for food industrial uses, for instance, in gasified beverages or cokes and yogurts, neither *o*-PA nor citric acid need to be removed from the sugar

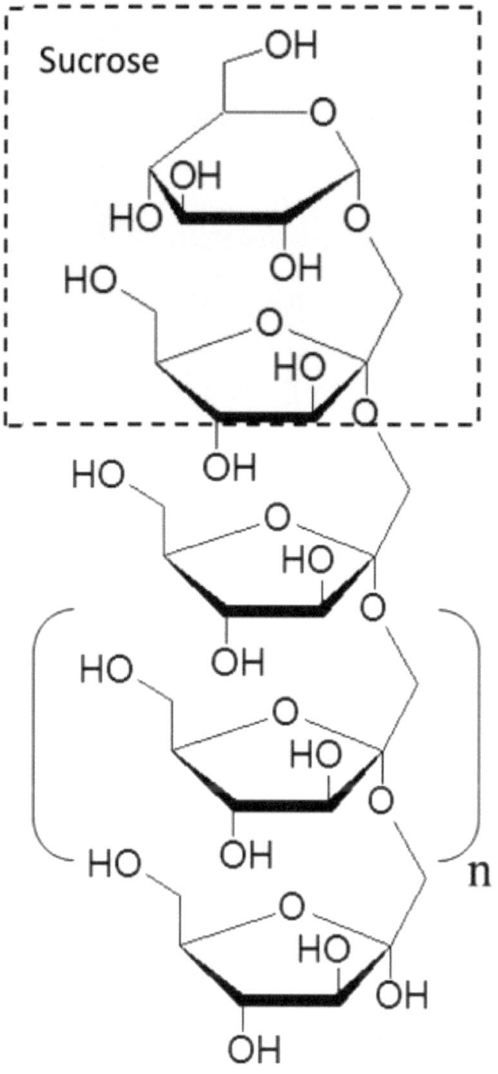

Figure 6. Simplified chemical structure of inulin, a β-linked polyfructofuranose.

partial hydrolysates, provided they are conveniently and partially neutralized to slightly less acid conditions (e.g., pH 3.5–4) with any desired bases (ammonia, lime, magnesium oxide, or even soda or potash). FOS, as extensively reported in the literature, are the most well-characterized nutraceuticals and particularly useful, when included in the human diet, blocking the appearance of colon tumors [38, 39]. **Figure 10** is a simplified summary of the factorial design leading to the best hydrolytic condition for the production of FOS from inulin. Herein, FOS are adopted as a family of the simpler fructobiose till larger oligosaccharides, DP = 17 or 18 and so on, and not simply the usual family of smaller FOS like kestose and neokestose mentioned in the literature and enzymatically obtained from transfructosylation of sucrose in a more expensive technology [40].

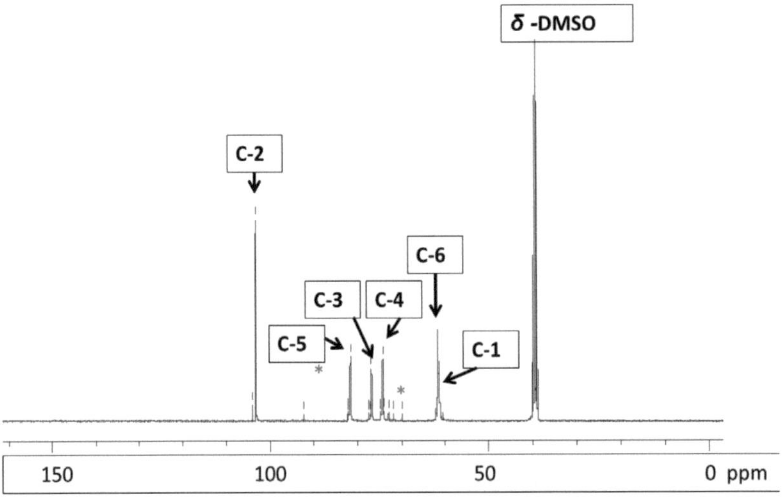

Figure 7. ^{13}C nuclear magnetic resonance spectrum of the purified inulin from *Dahlia* roots. Note: (*) minor signals arising from the single glucopyranose unit. Warm dimethyl sulfoxide (DMSO) as inulin solvent and ppm marker.

The phosphoric hydrolysate from inulin, optimized for FOS production, can be converted into short chain fatty acids (SCFAs), the real derivatives from FOS for the prevention of colonic tumors [41, 42].

11.2. Starch

Starch, the main reserve polysaccharide of many plants, fruits, and some algae, is a combination of linear amylose (exclusively α-1,4-linked glucopyranosyl linked) with α-1,6-branched amylopectin. Both of these glycosidic linkage types are moderately resistant to acid hydrolysis. We have prepared cassava starch purified granules by washing freshly collected roots, peeling, and comminuting them in a Waring blender and filtering the tick cold suspension through a 4× cheese cloth. After resting overnight, the starch grains settled and were again washed with distilled water for further lyophilization. Heavy starch paste was then pretreated with thermopressurized diluted phosphoric under selected conditions. The routine hydrolysis conditions varied from 2.8 to 6.8 atm, and the residence time at the peak temperature was from 5 to 10 min.

We have found that for cassava (manioc; *Cassava esculenta)* starch complete monomerization, the ideal conditions of thermopressurized hydrolysis, was polysaccharide concentration till 40% (w/v, *o*-PA effective pH = 1.75, and thermopressurization a 159°C/5 atm for a couple of minutes). HMF generation from glucose being released as free monosaccharide though its triple dehydration is lower when using *o*-PA than those generated at the same pH and severity hydrolysis parameters either using hydrochloric or sulfuric acids. This is graphically shown in the following **Figure 8**.

In the industrial jargon, the brown and bitter byproduct obtained when using these stronger acids is designed as "mud" and its elimination requires the use of expensive resins. Obviously,

http://dx.doi.org/10.5772/intechopen.68236

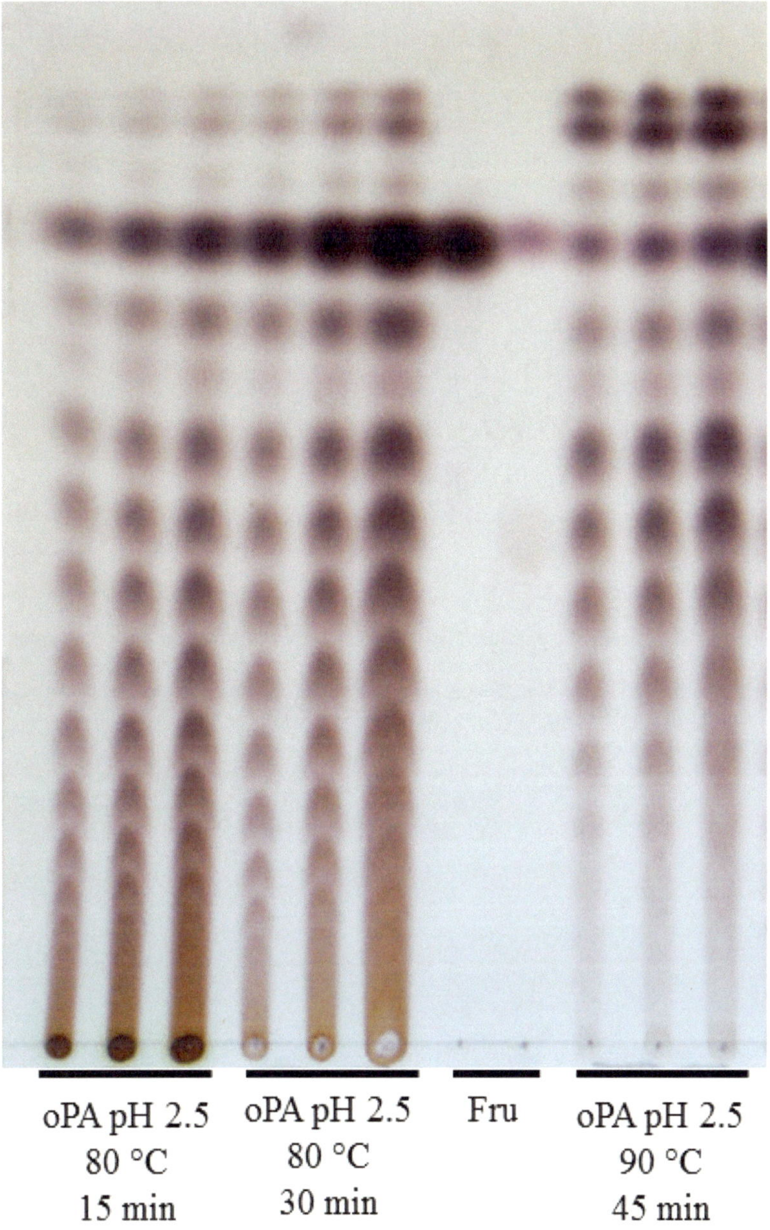

Figure 8. Thin-layer chromatographic analysis of the reaction products in the hydrolysis of inulin with phosphoric acid (*o*-PA) (pH 2.5) for 15, 30, and 45 minutes at 80°C. DP, degrees of depolymerization; FOS, fructo-oligosaccharides; Fru, fructose. Note: Silica gel 60 chromatoplates (Merck); isopropanol:ethyl acetate:water (5:2:1) was used as the mobile phase in three successive runs of 1/3, 2/3, and 3/3 of the front line, and hot 0.5 g% orcinol in sulfuric acid (10%) was used as the chromogenic agent).

o-PA technology allows modulation of the severity parameters to drive hydrolysis toward maltosaccharides instead of free glucose. It is worth to say that in comparison with inulin hydrolysis options (actually the industrial production of inulinases or fructofuranosidades

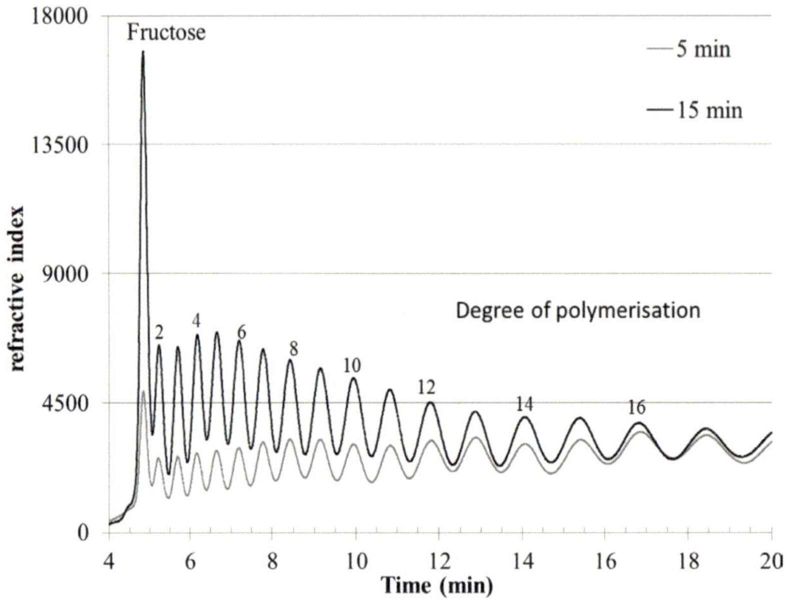

Figure 9. High performance liquid chromatography profile of fructooligosaccharides (FOS) from partial diluted phosphoric at pH 2.5 (at 85°C) hydrolysates of inulin. Note: High performance liquid chromatography analysis was performed in a Shimadzu LC-10 apparatus (Tokyo, Japan) consisting of a binary pump and a refractive index detector with a Spectra Amine column (200 × 4.6 mm, 5 μm) (Merck, Germany) and isocratic acetonitrile:water (7:3) at flow rate of 1.0 mL min^{-1} flow.

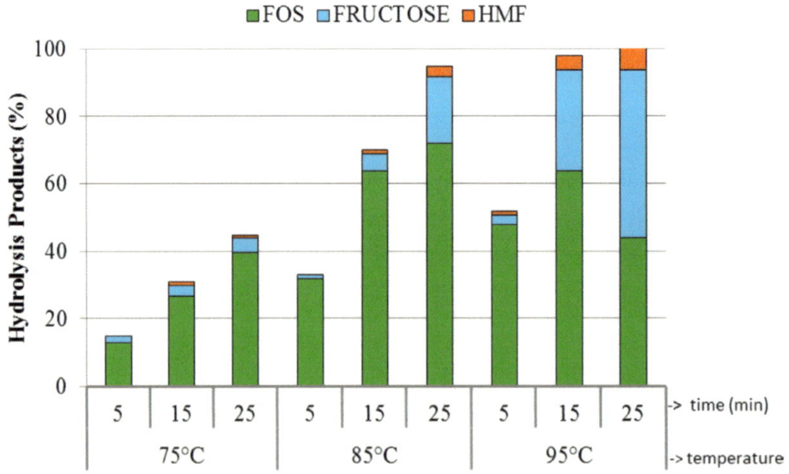

Figure 10. Comparative phosphoric acid-catalyzed production of fructooligosaccharides and fructose from inulin from *Dahlia* tubers inulin, exploring the kinetic variables time of reaction and temperature at a pH of 2.5.

seems to be in discontinuation), amylases still follow as good industrial tools (as is the case of corn starch conversion to glucose, in the USA) [43].

The quantitative distribution of released AMOS (AmyloOligosaccharides with DP from 2 to 10) following the severity parameter of H_3PO_4 mass as 32.4 or 132 mg g^{-1} starch and thermopressurization peak at 159°C (5 atm) were evaluated by densitometric measurement of spots intensity of a TLC plate (revealed with 0.5% orcinol in H_2SO_4:methanol (1:9); heating at 105°C for 5 min) in a Shimadzu equipment model flying spot CS-9301PC densitometer as shown in the following illustrations (**Figures 11** and **13**).

11.3. Ligno(hemi)cellulosics [L(h)C]

Ligno(hemi)cellulosics [L(h)C(s)] are by far more complex polymeric occurrences in phytobiomasses. In fact, cotton is a very peculiar natural form of pure cellulose since it is almost completely free of hemicelluloses and lignin (the case of cotton balls and not the whole plant). In average, L(h)Cs such as sugar cane bagasse, corn and other cereal straws, soya hulls, timber sawdust, and related materials are a tightly interbonded native architecture of cellulose:hemicellulose:lignin in an average proportion of 50:30:10 with lesser amounts of pectin, other minor polysaccharide, protein, salts, and orgasolvent-soluble extractives.

The two main distinctive features among L(h)Cs are the monolignol units in lignin (coumaryl, syringyl, and guayacyl) and the particular structure of hemicelluloses moieties as heteroxylans in hardwoods and grasses (angiosperms; monocotiledons) and heteromannans in softwoods (gymnosperms), although a minor amount of xylan is also present in conifer trees. The order of decreasing resistance to acid hydrolysis is: cellulose > mannans > xylan. It is also worth mentioning that xylan have other constituents (arabinose and (4-O-methyl)-glucuronic acid) are single branches (besides some O-acetyl groups in part of the basic xylosyl units), while the presence of glucose in conifer mannans is an intercalation between the basic mannobiosyl or mannotriosyl linear units. More, xylan branches are α-glycosidically linked, whereas glucosyl units of mannans are β-linked. In another words, L(h)Cs from hardwoods and grasses (e.g., cana bagasse) may be treated with thermopressurized phosphoric acid with two completely different purposes: (a) depolymerization (complete till free of xylose or partial till XOS – XyloOligoSaccharides); (b) almost complete lignin solubilization in the same phosphoric hydrolysate; and (c) labilization of the native cellulose component toward a new form (probably less crystalline) which is more prone to the action of cellulolytic enzymes.

A first report on phosphoric acid combined to other mineral acids for the depolymerization of oat straw came from the United States Department of Agriculture (USDA) researchers. The bioconversion of the released free sugars to grow less usual yeast, like *Aureobasidium pullulans*, and then using the hydrolysates as feeder for a field rodent [44]. Followed some years later, by our pioneering work with very diluted aqueous and thermopressurized phosphoric acid to monomerize cane and sorghum bagasse, to render the residual cellulose more labile to celluloses and hence its hydrolysis to free glucose to the final bioconversion to (bio) ethanol [45].

Here are plenty of chromatographic illustrations of factorial designs in order to optimize the hemicellulose hydrolyses of polysaccharides; the native models are sorghum starch, wheat hulls, and sugar cane bagasse. These illustrations correspond to **Figure 11**.

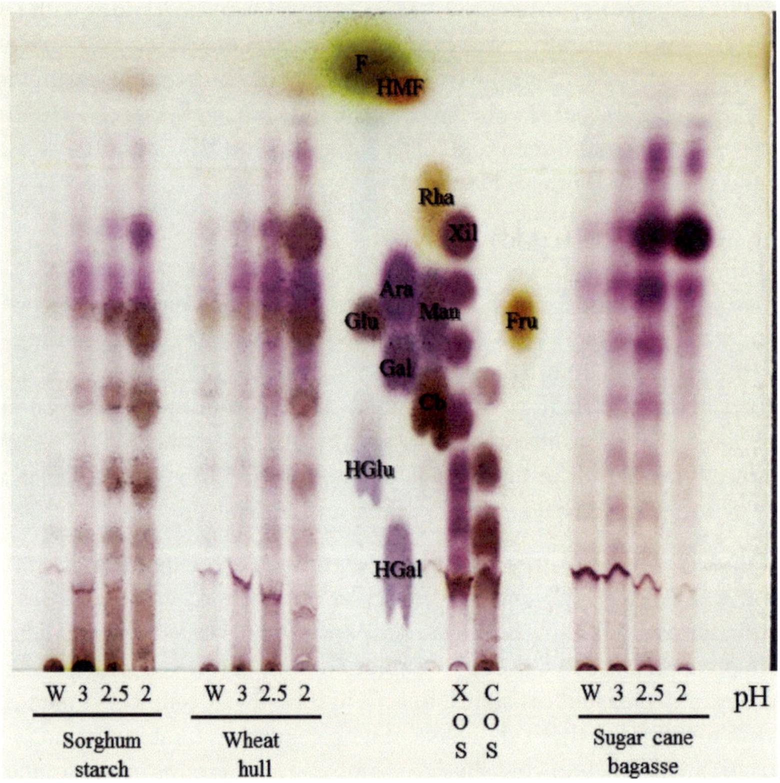

Figure 11. Thin-layer chromatographic for the progressive hydrolytic fragmentation of sorghum starch, wheat hulls, and sugar cane bagasse with pressure of 6.0 atm (165°C) and with water (w) and *o*-phosphoric acid pH of 2.0–3.0. Note: Center: standards of furfural (F), xylose (Xil), glucose (Glu), hydroxymethylfurfural (HMF), arabinose (Ara), rhamnose (Rha), fructose (Fru), galactose (Gal), mannose (Man), cellobiose (Cb), glucuronic acid (HGlu), galacturonic acid (HGal), cello oligosaccharides (COS) and xylooligosaccharides (XOS).

Formerly, FOS and later XOS, MOS, and GOS were reported as nutraceutical oligosaccharides deserving a better industrial and medical exploration. Some polysaccharides such as inulin and resistant starches are designed as functional foods or those who brings to men and animals benefits other than the simple nutrition. Nutraceutical is any food additive that corresponds to a concentrated form of a determined functional food. As a simple example, if carrot is considered a functional food, thanks to its antioxidant and free radicals fighter, β-carotene, its safe organosolvent extract (e.g., hot ethanol) may be considered the respective nutraceutical extract. *Bifidobacterium animalis* and *Lactobacillus casei* are two good examples of colonic beneficial bacterial microflora whose growth can be bioassayed for the appropriateness of nutraceutical oligosaccharides obtained from cane bagasse or pine sawdust. **Figure 12** shows the selected probiotic growth results.

Let us emphasize on the probable advantageous features of phosphoric acid as an ideal catalyst for phytobiomass polysaccharides depolymerization and further bioconversion to biofu-

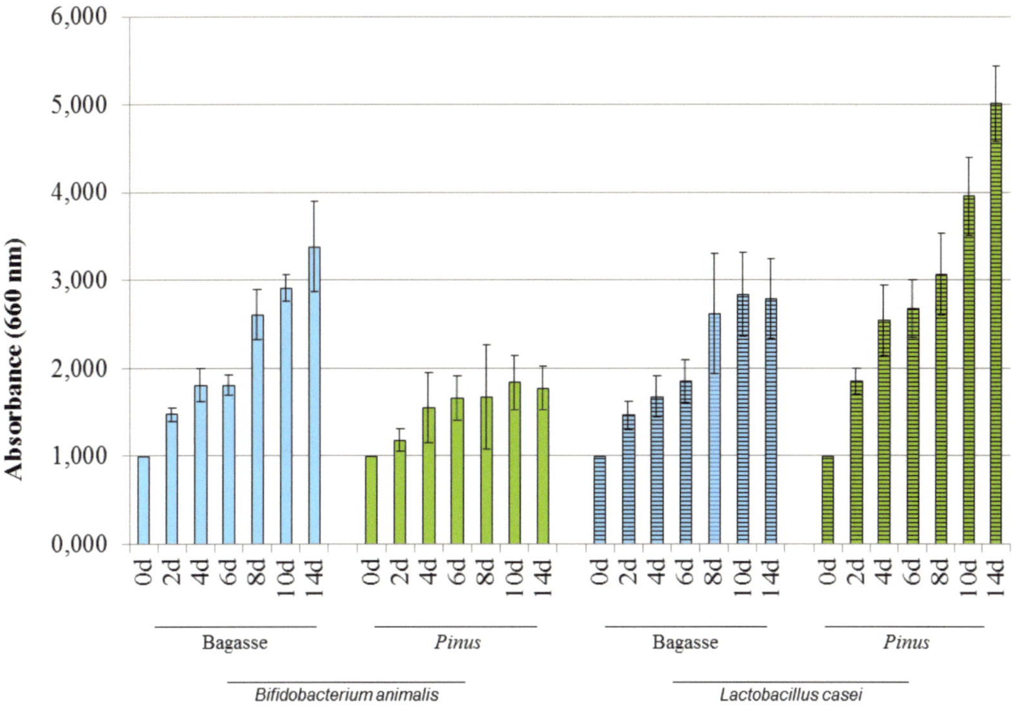

Figure 12. Growth of probiotic *Bifidobacterium* and *Lactobacillus* measured by spectrophotometric analysis on nutraceutical or prebiotic xylo- or mannooligosaccharides.

els and other industrial goods as compared to the stronger mineral acids. The hydrolysates resulting from the latter asks for desalting using expensive resins since chloride residual anions are often inhibitory to yeasts and other useful fermentation of microorganisms. Sulfate, if not so, is also less usefulness despite a minimum amount may be turned to the sulfur aminoacid methionine, although there are many steps of anabolism. Conversely, phosphoric may be maintained in the free sugar or oligosaccharide hydrolysates since together the latter and after partial neutralization with ammonia (or any other desirable base) it composes the usual and mandatory triplet for any industrial fermentation: free sugars and N and P sources.

Figure 13 assembles a comparative performance of diluted heated phosphoric acid, in this case through a simplified procedure, namely, an oven in Teflon-lined lid vials, for the partial (even total as well, if desirable) hydrolysis (oligosaccharidogenic effect) of corn cob xylan, *Dahlia* tubercles inulin and potato starch. All three oligosaccharidic families (OS, FOS, and MaltOS), then just following a partial neutralization till pH 4–5 with ammonia or soda may be used as C source for the growth of colon-beneficial bacteria such as *Lactobacillus* and *Bifidobacterium* spp., thus acting as nutraceutical oligosaccharides or symbiotics if as result of the blend of reagent (prebiotics) and products (probiotics).

Maltosaccharides are important feedstuff for the formulation of infant food. They can be easily produced using *o*-PA-catalyzed starch partial depolymerization (**Figure 14**). From the nutri-

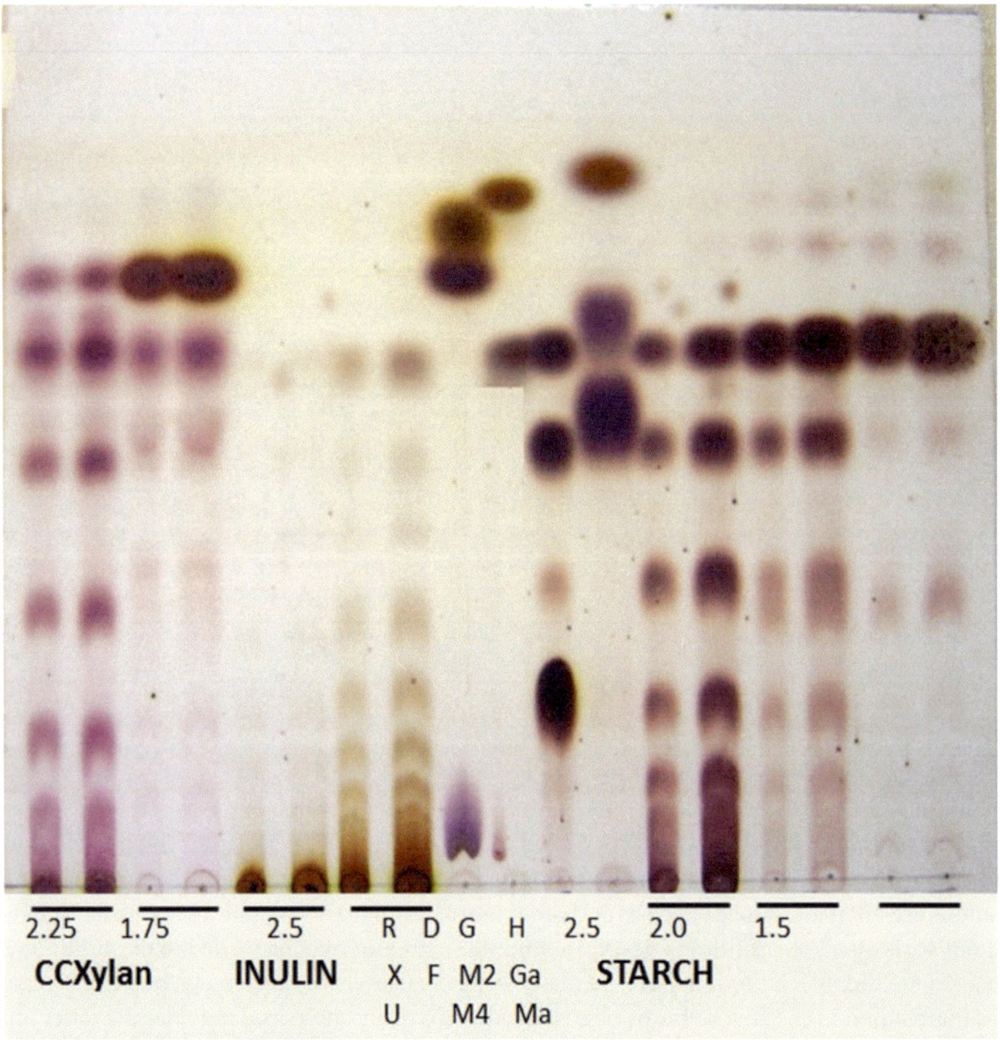

Figure 13. Thin-layer chromatographic of partial heated diluted phosphoric acid hydrolyses of corn cob Xylan, dahlia tubercles Inulin, and potato starch. Note: Numbers = hydrolyses pH at 160°C for 5 min for Xylan and Starch or at 80°C for Inulin (hydrolysates neutralized with $CaCO_3$); Standard: R, X, U = rhamnose, xylose and glucuronic acid; D, F = difructose III anhydride and fructose; G, M2, M4 = glucose, maltose and maltotetraose (with traces of maltotriose); H, Ga, Ma = hydroxymethylfurfural, galactose and mannose. Eluent = isopropylic alcohol:ethyl acetate:water (7:1:2).

tional standpoint, the neutralized catalyst with ammonia or alkaline bases can add important sources of P and N, for instance.

The destination of total or partial *o*-PA-hydrolyzed polysaccharides, depending on its source and hence carbohydrate composition, may be fermentation to biofuels (bioethanol or

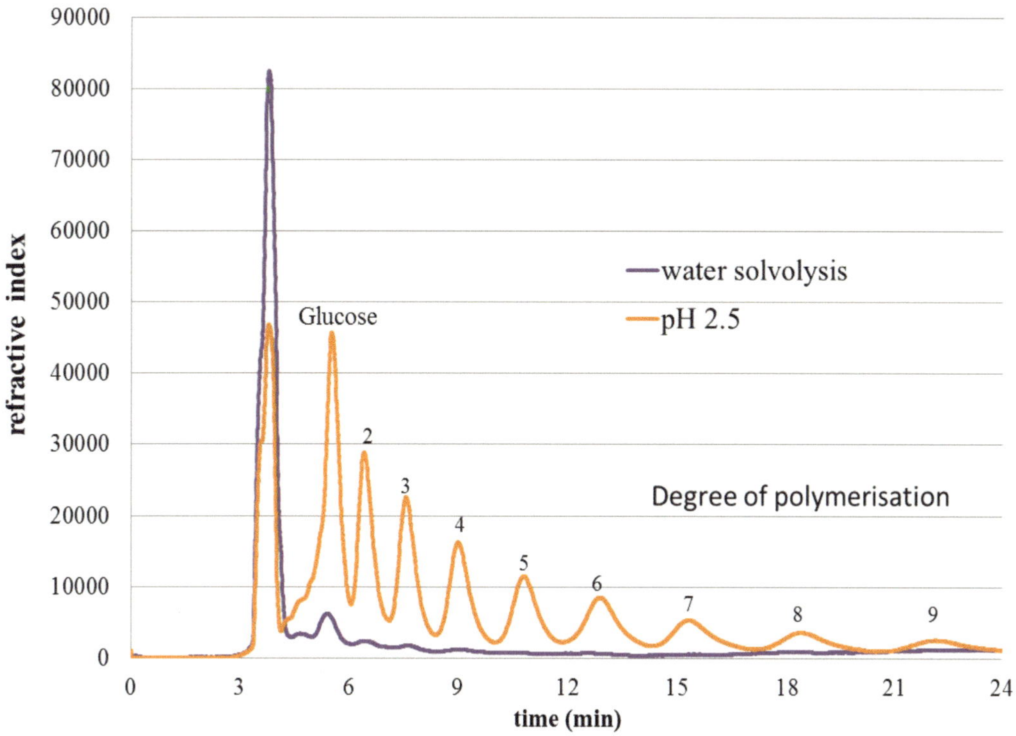

Figure 14. Oligosaccharide profiles of depolymerization of sorghum starch: comparative efficiencies of solvolysis (water) and *o*-PA. Note: 2–9 denotes for maltooligosaccharides with DP from maltose (2) till maltononaose (9). High performance liquid chromatography analysis were performed in a Shimadzu LC-10 apparatus (Tokyo, Japan) consisting of a binary pump and a refractive index detector with a Spectra Amine column (200 × 4.6 mm, 5 μm) (Merck, Germany) and isocratic acetonitrile:water (67:33) pH 9.0 (1.0 mL min^{-1} flow).

second-generation ethanol with *Saccharomyces cerevisiae* whenever free glucose from starch or fructose from inulin) or alternatively from pentoses and C5-oligosaccharides with other yeast such as the genders *Pichia, Candida,* and *Spathaspora* or generation of probiotics biomass (*Lactobacillus* and/or *Bifidobacterium* spp.) or even symbiotics (probiotics + prebiotics such as nutraceutical oligosaccharides) whenever the partial hydrolysates are richer in the appropriated oligosaccharides [46, 47].

For instance, soya husks hydrolyzed with pH 2.5 *o*-PA at 7 atm (171°C) released xylose a xylo-oligosaccharides as the major hydrolytic products, then converted to ethanol by the yeast, a known C-5 sugar ethanologenic fermenter (**Figure 15**).

Another example is the production of short chain fatty acids (SCFAs) by the beneficial enterobacterium *Bifidobacterium* growing in *o*-PA-partially hydrolyzed cell wall glycans from the cyanobacterium *Arthrospira* (formerly, *Spirulina*) (**Figure 16**). SCFAs are responsible for the colonic pH drop in turn inhibiting the growth of harmful coliforms, these implicated in the generation of aberrant crypta and later tumors [48].

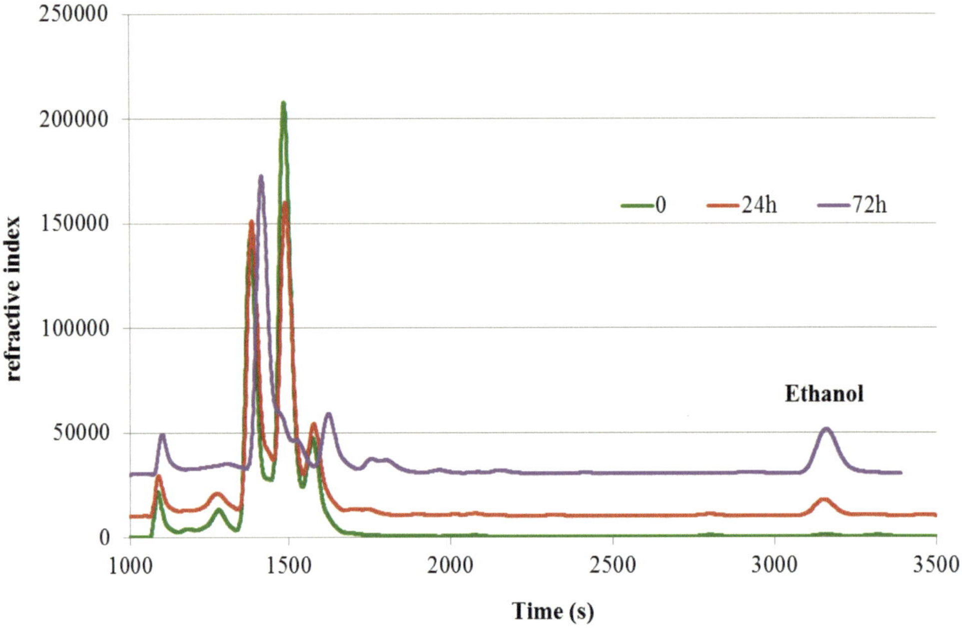

Figure 15. The progressive generation of ethanol from the yeast *Pichia stipitis* growing on *o*-PA-hydrolysed soya husks between 24 and 72 h of culture at 30°C. Note: High performance liquid chromatography analysis were performed in a Shimadzu LC-10 apparatus (Tokyo, Japan) consisting of a binary pump and a refractive index detector with a Rezex ROA column (300 × 7.8 mm, 8 μm) (Phenomenex, USA) and isocratic 8 mmol L^{-1} H_2SO_4 (0.5 mL min^{-1} flow).

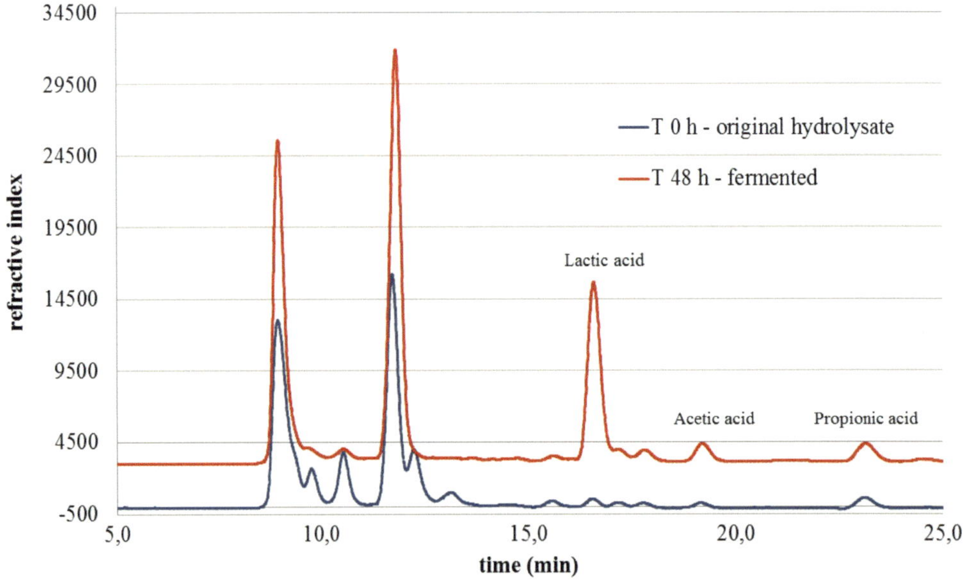

Figure 16. Time course of short chain fatty acids (SCFAs) production from the *o*-PA-hydrolyzed *Arthrospira platensis* (cyanobacterium) cell wall polysaccharides after incubation of the hydrolysate with *Bifidobacterium animalis*. Note: High performance liquid chromatography analysis was performed in a Shimadzu LC-10 apparatus (Tokyo, Japan) consisting of a binary pump and a refractive index detector with a Rezex ROA column (300 × 7.8 mm, 8 μm) (Phenomenex, USA) and isocratic 8 mmol L^{-1} H_2SO_4 (0.5 mL min^{-1} flow).

12. Phosphate analytical determination

There are a plenty of analytical methods for phosphate spectrophotometric determination. Examples are: molybdate/hydrazine [49], perchloric acid/molybdate/ascorbic acid (specially on phospholipids) [50], molybdate/Triton X-100 [51, 52], molybdate/quinolone [53], and molybdate/vanadate, too.

A comfortable and precise chromatographic determination is the utilization of ion chromatography that allows the parallel quantitation of other organic acids such as citric acid since both are widely used acidulants in the food and beverage industry [54].

13. Special applications of phosphoric acid and their derivatives

From the last century till recently, many novel technological applications of phosphoric acid came to light. A few examples are:

[A] An innovative chiral center was built on phosphoric acid by the insertion of two acidic phenolic groups. The compound acted as an efficient catalyst in the enantioselective alkylation of enones and indoles through the Friedel-Crafts methodology [55].

[B] 3-3-pentadecylphenylphosphoric acid (PDPPA), inserted with a side chain of a long hydrophobic hydrocarbon, as a doping for polyaniline and increasing its solubility in organosolvents, the plasticity for flexible films and allowing 1.8 S cm^{-1} of high conductivity [56].

[C] Cyclic ester amides of phosphoric acid (e.g., from N,N-bis-(B-chloroerhyl)-0,0-ethylene) were synthesized and proposed as favorable therapeutic cytostatics in the treatment of tumors [57].

[D] L-NASPA, N-palmitoyl-L-serine-phosphoric acid found its application as inhibitor of lysophosphatidyl receptors and affecting Ca^{2+} concentration in human sarcoma MG63 cells, thus indicating potential application in therapy [58].

Most of the applications are concerned, more recently, with improved fuel cells.

[E] In order to increase temperature resistance, a thermostable HT-PEMFC was built on polybenzimidazole previously doped with phosphoric acid. A range of 0.06–0.12 S cm^{-1} was achieved [59, 60].

[F] Fuel cells built on a composite of phosphoric acid and quaternary 1,4-diazabicyclo-[2.2.2]-octane (DABCO) polysulfone. Low and high degree of substitution led to the respective conductivities of 0.064 and 0.12 S cm^{-1} [61].

[G] Activity of phosphoric acid as a fuel cell electrolyte allowed operation at a higher temperature without loss of conductivity or circuit voltage. The novel membrane is based in the coordination of PO_4—groups with silicon which allowed power density of 184 mW cm^{-2} at 226°C with an H_2 flow rate of 4.1 mL min^{-1} [62].

[H] The direct oxidation of methanol in full cells (DMFC) was achieved in membranes made by grafting epoxy groups through PGMA, poly(glycidylmethacrylate). Imides were then generated therefrom and synthesis cycle was closed with phosphoric acid doping. The product was said to be of lower cost and with a performance superior to Naphion 117 [63].

14. Advertisement for readers

For those who wish a sharp view on recent advances on phosphoric acid, phosphates, and agricultural aspects of their applications and innovations, the recommendations are the annual series of "SYMPHOS" (International Symposium on Innovation and Technology in the Phosphate Industry) whose usual venue is Marrakesh, Morocoo. For instance, the 2015 event offered "Jacobs® New Process for Removing Iron from Phosphoric Acid" [64]; "Thickening, filtration and clarifying in Phosphoric Acid industry" [65]; "Comparison of different ways of desulfatation used in OCP phosphoric acid plants," (H. Mourchid, Responsible of Phosphoric Production, Maroc Phosphore Safi, OCP S.A., Morocco).

The 2013 edition of a review also encompasses a deeper view of advances experienced in both research and application of phosphates [66].

Beyond the scope of this chapter is the mention of worldwide generation of patent request in the subject, an interest which may be satisfied though quick searches on the recommended sites such as USPTO (USA), EPO (European Patent Office), and so on, all of them recorded at World Intellectual Property Organization (WIPO) and its directory.

Author details

José Domingos Fontana*, Marcela Tiboni and Heidegrid Siebert Koop

*Address all correspondence to: anatnof2012@gmail.com

Academic Department of Chemistry and Biology, Graduation Program on Environmental Science & Technology/UTFPR, LQBB, Biomassses Chemo/Biotechnology Laboratory/DAQBI, PPGCTA, Federal Technological University of Parana, Curitiba, PR, Brazil

References

[1] Nautilus. Phosphorus [Internet]. 1999. Avaiable from: http://nautilus.fis.uc.pt/st2.5/index-en.html [accessed: 9 September 2016]

[2] McKetta J.J., Cunningham W.A. 36 – Phosphorus to pipeline failure subsidence strain. In: Encyclopedia of Chemical Processing and Design vol. 36 - John J. McKetta Jr (editor). New York, USA: Marcel Dekker INC; 1991

[3] Readman JB. Process of Obtaining Phosphorus. USA Patent US417943 A. 1889.

[4] Sanderson RT. Chemical Element: Phosphorus (P) [Internet]. 2009. Avaiable from: https://global.britannica.com/science/phosphorus-chemical-element [accessed: 1 September 2016]

[5] Raymond Harrison MB. 2009. Bilateral lens opacities: Associated with use of Di-isopropyl fluorophosphate eyedrops. American Journal of Ophthalmology. **50**:153-154. DOI: 10.1016/0002-9394(60)90852-7

[6] Millard CB, Kryger G, Ordentlich A, Greenblatt HM, Harel M, Raves ML, Segall Y, Barak D, Shafferman A, Silman I, Sussman JL. 1999. Crystal structures of aged phosphonylated acetylcholinesterase: Nerve agent reaction products at the atomic level. Biochemistry. **38**:7032-7039. DOI: 10.1021/bi982678l

[7] CBN. 1978. Nomenclature of phosphorus-containing compounds of biochemical importance. (Recommendations 1976). IUPAC-IUB Commission on Biochemical Nomenclature. Biochemical Journal. **171**:1-19. PMC1184129

[8] York. Basic Chemicals: Phosphorus [Internet]. 2013. Avaiable from: http://www.essentialchemicalindustry.org/chemicals/phosphorus.html [accessed: 2 September 2016]

[9] Takeshita EM, Castro LP, Sassaki KT, Delbem ACB. 2009. In vitro evaluation of dentifrice with low fluoride content supplemented with Trimetaphosphate. Caries Research. **43**:50-56. DOI: 10.1159/000196507

[10] Christoffersen J, Christoffersen MR, Kibalczyc W, Perdok WG. 1986 Kinetics of dissolution and growth of calcium fluoride and effects of phosphate. Acta Odontologica Scandinavica. **46**:325-336. DOI: 10.3109/00016358809004784

[11] Delbem ACB, Bergamaschi M, Rodrigues E, Sassaki KT, Vieira AEDM, Missel EMC. 2012. Anticaries effect of dentifrices with calcium citrate and sodium trimetaphosphate. Journal of Applied Oral Science. **20**:94-98. DOI: http://dx.doi.org/10.1590/S1678-77572012000100017

[12] Zero Domenick T. 2006. Dentifrices, mouthwashes, and remineralization/caries arrestment strategies. Biotechnology and Biomaterials to Reduce Caries Epidemic Oral Health. **6**:S9-S. 10.1186/1472-6831-6-s1-s9

[13] Totten GE, Negri VJd. 2011. Handbook of Hydraulic Fluid Technology. 2nd ed. CRC Press; Boca Raton, Florida, USA.

[14] Svara J, Weferling N, Hofmann T. 2000. Phosphorus compounds, organic. In: editors. Ullmann's Encyclopedia of Industrial Chemistry. Editor: Wolfgang Gerhartz. Wiley-VCH Verlag GmbH & Co.; New York, USA. KGaA; 10.1002/14356007.a19_545.pub2

[15] Levchik SV. Weil ED. 2006. A review of recent progress in phosphorus-based flame retardants. Journal of Fire Sciences. **24**:345-364. DOI: 10.1177/0734904106068426

[16] Gilmour R. Phosphoric Acid: Purification, Uses, Technology and Economics. Boca Raton, FL, USA: CRC Press; 2013

[17] TOXNET. Substance Name: Aluminum Phosphate [Internet]. 1965. Avaiable from: https://chem.sis.nlm.nih.gov/chemidplus/rn/7784-30-7 [accessed: 2 September 2016]

[18] Monsanto. More about Phosphorus [Internet]. 2015. Avaiable from: http://www.monsanto.com/soda-springs/pages/more-about-phosphorus.aspx [accessed: 5 September 2016]

[19] Chabra N. Metabolism of glycine (Part 2) – Metabolic role and clinical significance" in: Biochemistry for Medics [Internet]. 2014. Avaiable from: http://www.namrata.co/metabolism-of-glycine-part-2-metabolic-role-and-clinical-significance/[accessed: 30 May 2016]

[20] KierL, KirklandD. 2013. Review of genotoxicity studies of glyphosate and glyphosate-based formulations. Critical Reviews in Toxicology. **43**:283-315. DOI: 10.3109/10408444.2013.770820

[21] Schinasi L, Leon M. 2014. Non-Hodgkin lymphoma and occupational exposure to agricultural pesticide chemical groups and active ingredients: a systematic review and meta-analysis. International Journal of Environmental Research and Public Health. **11**:4449-4527. DOI: 10.3390/ijerph110404449

[22] Cressey D. Widely used herbicide linked to cancer. Nature News (24 March 2015) doi:10.1038/nature.2015.17181

[23] Kazemi M, Tahmasbi A, Valizadeh R, Naserian AA, Soni A. Organophosphate pesticides: A general review. Agricultural Science Research Journals. **2**:512-522

[24] Hall J, Wickenden J, Yau K. Biochemical Conjugation of Pesticides in Plants and Microorganisms: An Overview of Similarities and Divergences. Washington, USA: American Chemical Society; 2001

[25] FAS. White Phosphorus Fact Sheet [Internet]. 2013. Avaiable from: http://fas.org/programs/bio/factsheets/whitephosphorus.html [accessed: 05 September 2016]

[26] Mochida I, Korai Y, Shirahama M, Kawano S, Hada T, Seo Y, Yoshikawa M. Yasutake A. 2000. Removal of SOx and NOx over activated carbon fibers. Carbon. **38**:227-239. DOI: 10.1016/S0008-6223(99)00179-7

[27] Ichcho S, Khouya E, Fakhi S, Ezzine M, Hannachem H, Pallier R, Naslain R. 2005. Influence of the experimental conditions on porosity and structure of adsorbents elaborated from Moroccan oil shale of Timahdit by chemical activation. Journal of Hazardous Materials. **118**:45-51. DOI: http://dx.doi.org/10.1016/j.jhazmat.2004.10.009 [http://www.seaweed.ie/algae/redtides.php [accessed:] 5.July.2016

[28] Suarez-Garcia F, Martınez-Alonso A, Tascon J. Activated carbon fibers from Nomex by chemical activation with phosphoric acid. Carbon. **42**:1419-1426. DOI: 10.1016/j.carbon.2003.11.011

[29] Guiry M. The Seaweed Site: Information on Marine Algae [Internet]. 2016. Avaiable from: http://www.seaweed.ie/algae/redtides.php [accessed: October 8, 2016]

[30] Weaver K, Kim H, Sun J, MacFarlane D, Elliott G. 2005. Cyto-toxicity and biocompatibility of a family of choline phosphate ionic liquids designed for pharmaceutical applications. Green Chemistry. **12**:507-513. DOI: 10.1039/B918726J

[31] Rowe RC, Sheskey PJ, Quinn ME. Handbook of Pharmaceutical Excipients. 6th ed. Italy: RPS Publishing; 2009

[32] Schmidt PC, Herzog R. Calcium phosphates in pharmaceutical tableting. 1993. 2. Comparison of tableting properties. International Journal of Clinical Pharmacy. **15**:116-122. DOI: 10.1007/BF02113939

[33] Melo MAS, Cheng L, Zhang K, Weir MD, Zhou X, Bai Y, Rodrigues LKA, Xu HHK. Chapter 6. Novel Nanostructured Bioactive Restorative Materials for Dental Applications. CRC Press; Boca Rato, Florida, USA. 2015. DOI: 10.1201/b18654-8

[34] Dawson RMC, Elliott DC, Elliott WH, Jones KM. Data for Biochemical Research. 3 ed. New York, USA: Clerendon Press, Oxford Science Publications; 1989

[35] Pazianas M, Miller P, Blumentals WA, Bernal M, Kothawala P. 2007. A review of the literature on osteonecrosis of the jaw in patients with osteoporosis treated with oral bisphosphonates: Prevalence, risk Factors, and clinical characteristics. Clinical Therapeutics, **29**:1548-1558. DOI: 10.1016/j.clinthera.2007.08.008

[36] Shinkai I, Ohta Y. 1996. New drugs-reports of new drugs recently approved by the FDA. Alendronate. Bioorganic & Medicinal Chemistry. **4**:3-4. DOI:10.1016/0968-0896(96)00042-9

[37] Black DM, Cummings SR, Karpf DB, Cauley JA, Thompson DE, Nevitt MC, Bauer DC, Genant HK, Haskell WL, Marcus R, Ott SM, Torner JC, Quandt SA, Reiss TF, Ensrud KE. 1996. Randomised trial of effect of alendronate on risk of fracture in women with existing vertebral fractures. The Lancet. **348**:1535-1541. DOI: 10.1016/s0140-6736(96)07088-2

[38] Singh J, Rivenson A, Tomita M, Shimamura S, Ishibashi N, Reddy BS. 1997. *Bifidobacterium longum*, a lactic acid-producing intestinal bacterium inhibits colon cancer and modulates the intermediate biomarkers of colon carcinogenesis. Carcinogenesis. **18**:833-841. DOI: 10.1093/carcin/18.4.833

[39] Challa A, Rao DR, Chawan CB, Shackelford L. 1997. Bifidobacterium longum and lactulose suppress azoxymethane-induced colonic aberrant crypt foci in rats. Carcinogenesis. **18**:517-521. DOI: 10.1093/carcin/18.3.517

[40] Fontana JD, Grzybowski A, Tiboni M, Passos M. 2011. Fructo-Oligosaccharide production from inulin through partial citric or phosphoric acid hydrolyses. Journal of Medicinal Food. **14**:1425-1430

[41] Gamet L, Daviaud D, Denis-Pouxviel C, Remesy C, Murat J. 1992. Effects of short-chain fatty acids on growth and differentiation of the human colon-cancer cell line HT29. International Journal of Cancer. **52**:286-289. DOI: 10.1002/ijc.2910520222

[42] Gibson GR, Roberfroid MB. 1995. Dietary modulation of the human colonic microbiota: Introducing the concept of prebiotics. The Journal of Nutrition. **125**:1401-1412. DOI: 10.1079/NRR200479

[43] Fontana JD, Passos M, Baron M, Mendes SV, Ramos LP. 2001. Cassava starch Maltodextrinization/monomerization through thermopressurized aqueous phosphoric acid hydrolysis†. Applied Biochemistry and Biotechnology. **91-93**:469-478

[44] Israilides CJ, Grant GA, Han YW. 1978. Sugar level, fermentability, and acceptability of straw treated with different acids. Applied and Environmental Microbiology. **36**:43-46

[45] Fontana JD, Correa JBC, Duarte JH, Blumel M. 1984. Aqueous phosphoric acid hydrolysis of sugar cane and sorghum bagasses. Biotechnology and Bioengineering Symposium. **14**:175-184

[46] Cunha-Pereira Fd, Hickert LR, Sehnem NT, de Souza-Cruz PB, Rosa CA, Ayub MAZ. 2011. Conversion of sugars present in rice hull hydrolysates into ethanol by Spathaspora arborariae, saccharomyces cerevisiae, and their co-fermentations. Bioresource Technology. **102**:4218-4225. DOI: http://dx.doi.org/10.1016/j.biortech.2010.12.060

[47] Su Y-K, Willis LB, Jeffries TW. 2015. Effects of aeration on growth, ethanol and polyol accumulation by Spathaspora passalidarum NRRL Y-27907 and scheffersomyces stipitis NRRL Y-7124. Biotechnology and Bioengineering. **112**:457-469. DOI: 10.1002/bit.25445

[48] Hsu C-K, Liao J-W, Chung Y-C, Hsieh C-P, Chan Y-C. 2004. Xylooligosaccharides and fructooligosaccharides affect the intestinal microbiota and precancerous colonic lesion development in rats. The Journal of Nutrition. **134**:1523-1528

[49] Pradhan S, Pokhrel MR. 2013. Spectrophotometric determination of phosphate in sugarcane juice, fertilizer, detergent and water samples by molybdenum blue method. Scientific World. **11**:58-62. 10.3126/sw.v11i11.9139

[50] Rouser G, Fleischer S, Yamamoto A. Two dimensional thin layer chromatographic separation of polar lipids and determination of phospholipids by phosphorus analysis of spots. Lipids. **5**:494-496. 10.1007/bf02531316

[51] Dzimiri N, Almotrefi AA. Relationship between potassium concentration and inhibitory effects of β-adrenergic blockers on myocardial Na^+, K^+-ATPase. Drug Investigation. **4**:166-172. 10.1007/bf03258395

[52] Eibl H, Lands WEM. 1969. A new, sensitive determination of phosphate. Analytical Biochemistry. **30**:51-57. 10.1016/0003-2697(69)90372-8

[53] United States Department of Agriculture Food Safety and Inspection Service, Office of Public Health Science CLG-PHS1.01 Page 1 of 10 Title: Determination of Phosphate Revision: 01 - 08/10/2009 / Official Methods of Analysis of the Association of Official Analytical Chemists, 15th Edition, 969.31B. https://www.fsis.usda.gov/wps/wcm/connect/87172299-2fd3-41fd-aa3a-ca603ce7fcb6/CLG_PHS_1_01.pdf?MOD=AJPERES

[54] Dionex. Application Note 164 – Assay for Citrate and Phosphate in Pharmaceutical Formulations Using Ion Chromatography [Internet]. 2004. Avaiable from: http://www.dionex.com/en-us/webdocs/49183-AN164_LPN1643.pdf [accessed: 2 September 2016]

[55] Liu X-L, Yu Z-B, Pan B-W, Chen L, Feng T-T, Zhou Y. 2015. Synthesis of novel chiral phosphoric acid-bearing two acidic phenolic hydroxyl groups and its catalytic evaluation for

enantioselective Friedel-Crafts alkylation of indoles and enones. Journal of Heterocyclic Chemistry. **52**:628-634. DOI: 10.1002/jhet.1949

[56] Paul RK, Vijayanathan V, Pillai CKS. 1996. Melt/solution processable conducting polyaniline: doping studies with a novel phosphoric acid ester. Synthetic Metals. **104**:189-195. DOI: http://dx.doi.org/10.1016/S0379-6779(99)00063-6

[57] Bourseaux F, Arnold H, Brock N. Novel Cyclic Phosphoric Acid Ester Amides, and the Production Thereof. USA Patent US 3018302 A; 1962

[58] Jan CR, Lu YC, Jiann BP, Chang HT, Wang JL, Chen WC, Huang JK. 2003. Novel effect of N-palmitoyl-L-serine phosphoric acid on cytosolic Ca^{2+} levels in human osteoblasts. Pharmacology Toxicology. **93**:71-76. DOI: 10.1034/j.1600-0773.2003.930203.x

[59] Mack F, Aniol K, Ellwein C, Kerres J, Zeis R. Novel phosphoric acid-doped PBI-blends as membranes for high-temperature PEM fuel cells. Journal of Materials Chemistry A. **3**:10864-10874. DOI: 10.1039/c5ta01337b

[60] Yeager GW, Krishnan L, Early TA, Zhang TC, LaTorre MR. 2013. Novel Polybenzimidazole-Phosphoric acid membranes for fuel cell applications. Electronic Clearing System Transactions. **50**:1179-1191. 10.1149/05002.1179ecst

[61] Wang X, Xu C, Golding BT, Sadeghi M, Cao Y, Scott K. 2011. A novel phosphoric acid loaded quaternary 1,4-diazabicyclo-[2.2.2]-octane polysulfone membrane for intermediate temperature fuel cells. International Journal of Hydrogen Energy. **36**:8550-8556. DOI: http://dx.doi.org/10.1016/j.ijhydene.2011.03.143

[62] Ansari Y, Tucker TG, Angell CA. 2013. A novel, easily synthesized, anhydrous derivative of phosphoric acid for use in electrolyte with phosphoric acid-based fuel cells. Journal of Power Sources. **237**:47-51. DOI: http://dx.doi.org/10.1016/j.jpowsour.2013.03.003

[63] Abu-Saied MA, Elzatahry AA, El-Khatib KM, Hassan EA, El-Sabbah MM, Drioli E, Mohy Eldin MS. Preparation and characterization of novel grafted cellophane-phosphoric acid-doped membranes for proton exchange membrane fuel-cell applications. Journal of Applied Polymer Science. **123**:3710-3724. DOI: 10.1002/app.35048

[64] Hilakos SW. Jacobs' new process for removing iron from phosphoric acid. In: 39th Annual Clearwater Conference; Bartow, Florida, USA; 2015. Sponsored by AIChE) – American Institute of Chemical Engineers.

[65] Lake P, Cutler R, Gaspar A, Summerhays R. Thickening, Filtration and Clarifying in Phosphoric Acid industry WesTech Inc. Salt Lake City, UT, USA: Marrakesh; 2015

[66] Acton QA. Phosphatases—Advances in Research and Applications. Atlanta, USA: Scholarly Editions; 2013

Section 2

Industrial Processing

Chapter 4

Phosphoric Acid Industry: Problems and Solutions

Benjamín Valdez Salas, Michael Schorr Wiener and
Juan Ricardo Salinas Martinez

Additional information is available at the end of the chapter

http://dx.doi.org/10.5772/intechopen.70031

Abstract

Phosphoric acid (PA) is an important industrial chemical used as an intermediate in the fertilizer industry, for metal surface treatment in the metallurgical industry and as an additive in the food industry. The PA industry is spread out worldwide in Europe, Asia and America, including countries that operate phosphate rock (PR) mines and produce PA, phosphatic fertilizers and phosphate-based products.

Keywords: corrosion, phosphoric acid, phosphate rock, erosion-corrosion, WPA

1. Introduction

Phosphoric acid (PA) is an important industrial chemical used as an intermediate in the fertilizer industry, for metal surface treatment in the metallurgical industry and as an additive in the food industry. The PA industry is spread out worldwide in Europe, Asia and America, including countries that operate phosphate rock (PR) mines and produce PA, phosphatic fertilizers and phosphate-based products. The PR reserves; worldwide are shown in **Figure 1**.

Acids, such as phosphoric (H_3PO_4), sulfuric (H_2SO_4), nitric (HNO_3), hydrochloric (HCl) and acetic (CH_3COOH) acids, are broadly applied in many industries: chemical, fertilizers, mineral leaching, water purification, petroleum refining, food and metal production [1–3].

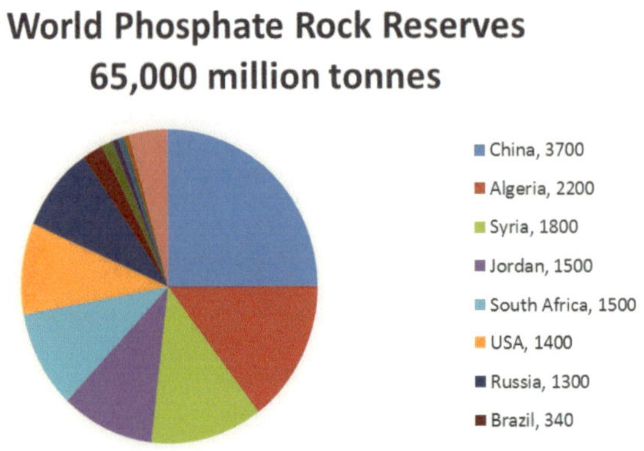

Figure 1. World phosphate rock reserves. Source: USSG mineral commodity summary 2011.

2. The world phosphate rock (PR) market

PR is the raw material for the production of the diverse types of PA. It is mined in several continents and countries, as recorded in **Table 1**.

Countries	Production (in '000 metric tons)
China	65,000
USA	26,100
Morocco and Western Sahara	26,000
Tunisia	10,000
Jordan	7600
Brazil	6000
Egypt	5500
Israel	5000
Australia	3000
Syria	2800
South Africa	2300
Algeria	2000
Togo	800
Canada	700
Senegal	650
Others Countries	9500

Source: USSG Mineral Commodity Summary 2011.

Table 1. World producers of phosphate rock.

http://dx.doi.org/10.5772/intechopen.70031

3. Phosphate rock: types and chemical composition

Apatite is the most important PR applied in the production of industrial and food grade PA. It is mainly of the fluoride and hydroxide type. Both rocks contain many impurities: fluoride (F^-), chloride (Cl^-), iron and aluminum oxides (Fe_2O_3, Al_2O_3); pyrites (ferrous sulfide, FeS) and fossilized organic matter. These interfere with production procedures, impair the PA quality and enhance corrosion and deterioration of metallic and plastic materials used for fabrication of equipment and rubber linings for reactor walls [4–6]. The chemical composition (in weight percentage) of PRs from different sources is given in **Table 2** [7].

4. Phosphoric acid: production processes and industrial plants

The main wet processes for PA production are:

a. The Wet process acid (WPA) uses sulfuric acid to convert the PR into PA and

b. The solvent extraction (SX) process, where the PR is acidulated with HCl and PA is separated applying an industrial alcohol as solvent.

4.1. The WPA consists of three main stages

Acidulation of PR by H_2SO_4; the overall reaction with fluoroapatite is usually expressed as:

$$Ca_{10}(PO_4)_6F_2 + 10H_2SO_4 \rightarrow 10CaSO_4 + 6H_3PO_4 + 2HF \quad (1)$$

The hydrogen fluoride reacts with any active silica present to form silicon tetrafluoride:

$$4HF + SiO_2 \rightarrow SiF_4 + 2H_2 \quad (2)$$

It volatilizes as such or hydrolyzes to fluorosilicic acid and forms silica deposits:

$$3SiF_4 + 2H_2O \rightarrow 2H_2SiF_6 + SiO_2 \quad (3)$$

During the acidulation, thick slurry is formed containing 30% of solid particles, mainly gypsum ($CaSO_4$) and unreacted PR components.

Filtration is performed to separate the solid particles from the filter acid, 30% w/w P_2O_5 (50% PA).

Concentration is performed by evaporation of the filter acid to merchant grade PA, 54% w/w P_2O_5 (70% PA).

The H_2SO_4 reacts with any calcium carbonate and heavy metal oxides present in most rocks:

$$CaCO_3 + H_2SO_4 \rightarrow CaSO_4 + CO_2 + H_2O \quad (4)$$

$$Fe_2O_3 + 3H_2SO_4 \rightarrow Fe_2(SO_4)_3 + 3H_2O \quad (5)$$

$$Al_2O_3 + 3H_2SO_4 \rightarrow Al_2(SO_4)_3 + 3H_2O \quad (6)$$

Constituent	Florida, USA	North Carolina, USA	Palfos, RSA	Pesca, Colombia	Hazara, Pakistan	Monte Fresca, Venezuela	Araxa, Brazil	Hidalgo, Mexico	Sahara, Morocco	Safi, Morocco	Ruseifa, Jordan	Oron, Israel
P_2O_5	31.2	29.7	39.9	20.5	28.5	34.18	35.5	43.3	34.2	32.4	33.4	29.8
CaO	45.0	47.4	–	29.0	41.9	42.30	47.3	46.3	50.3	49.9	51.0	51.0
Cl	0.05	0.015	–	0.001	0.03	–	0.001	0.02	0.02	0.02	4.2	0.03
F	3.60	3.53	2.35	2.0	2.92	2.94	2.54	–	3.8	4.1	4.9	3.8
SiO_2	9.48	1.73	1.0	39.2	23.2	10.29	0.41	2.8	–	2.85	0.2	0.68
Fe_2O_3	1.33	0.79	–	0.8	1.85	0.66	2.42	–	0.22	0.70	0.3	0.2
Al_2O_3	1.76	0.53	0.35	1.1	1.0	1.15	0.32	1.11	0.48	0.40	–	0.3
MgO	–	0.79	0.51	0.09	0.13	0.21	0.07	2.17	0.12	0.70	–	–
Na_2O	0.89	0.98	–	0.14	0.16	1.30	0.03	0.05	–	0.90	–	–
K_2O	0.11	0.17	–	0.14	0.31	0.18	0.10	–	–	0.10	4.5	–
CO_2	3.48	4.18	1.0	3.0	1.1	–	1.7	0.02	2.7	4.1	–	7.8
Organic C	2.18	1.38	–	0.3	0.18	–	<0.1	–	0.06	–	–	0.6
Total S	1.05	1.1	–	0.1	0.18	–	1.52	–	–	0.20	–	–

Table 2. Chemical composition of phosphate rocks.

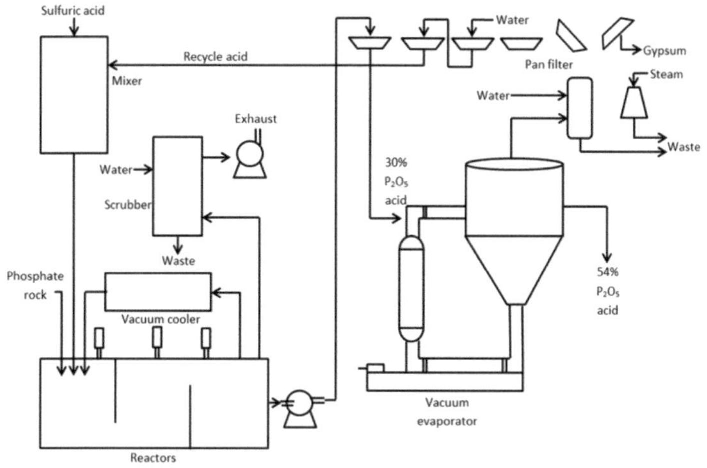

Figure 2. Typical wet process industrial plant.

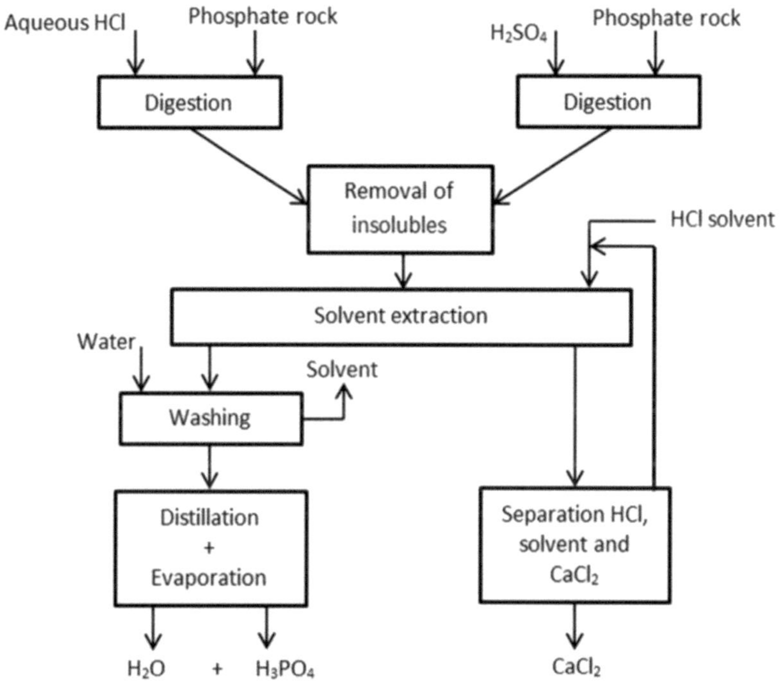

Figure 3. Diagram of solvent extraction process.

PRs contain chlorides, such as NaCl, which yields HCl:

$$2NaCl + H_2SO_4 \rightarrow Na_2SO_4 + 2HCl \quad (7)$$

The PR industry requires large amounts of water to wash its ore and to dissolve corrosive chlorides. These Negev desert waters were investigated to characterize their scaling, corrosion and fouling mechanisms, which affect the water system [8]. A critical unit is the huge reactor into which PR and H_2SO_4 are fed (**Figure 2**).

4.2. The solvent extraction process

PR is reacted with HCl to form an aqueous reaction mixture comprising PA and $CaCl_2$:

$$Ca_3(PO_4)_2 + 6\ HCl \rightarrow 2H_3PO_4 + 3\ CaCl_2 \qquad (8)$$

In this version, the slurries obtained by treatment with H_2SO_4 (wet process) and with HCl (SX process) are mixed, and subsequently, the PA is extracted with an aliphatic alcohol (**Figure 3**) [9–11].

5. Phosphate rock impurities: influence on production

PR is the principal source of dissolved and suspended impurities in wet process phosphoric acid (WPA). Other impurities, such as Cl^- may be introduced in process water, particularly brackish water. Sometimes contaminated H_2SO_4 obtained from the hydrometallurgy industry introduces additional impurities. The impurities impart undesired color and turbidity to the WPA and increase the corrosiveness of PA. Cl^- and F^- are particularly corrosive, and other impurities that affect corrosion are SiO_2, Al_2O_3, alkali metal salts, SO_4^-, S^-, organic matter and oxidizing agents. The corrosivity of a specific impurity depends on its chemical nature, the concentration of its active species, and its interaction with other acid constituents and with the surface of the specific metal [12, 13].

In addition to affecting corrosion, some impurities change the density and viscosity of the acid and form acid sludge and sediments. In contact with metallic surfaces, these sediments may influence the corrosion behavior of the metal by forming deposits that promote localized corrosion.

5.1. Influence of chloride ion

The chloride ion (Cl^-) in PA comes from phosphate ores in which it is present as an alkali metal chloride, such as NaCl. It may be removed by washing with fresh water, but it may also come from the wash water since seawater is sometimes used for washing the PR or forming rising tanks or cargo holds. The Cl^- content then can rise to a dangerous level. Chloride may also occur in the apatite itself, sometimes as a water-insoluble salt or as an oxychloride formed during calcination of the PR.

Chloride contamination may also result from handling and storing of raw materials. In many fertilizer plants, PR and muriate of potash (MOP, the most common source of potassium used in agriculture as potassium chloride) are unloaded with the same equipment and stored in bulk in the same building. Small amounts of MOP often are left in elevators and conveyors and may be mixed with the PR subsequently handled in the equipment.

The corrosivity of the halogen acids, HCl and HF, and of the halide ions Cl^- and F^- in strong mineral oxygen-acids, such as H_3PO_4 and H_2SO_4, is related to the physicochemical properties of the halogens, and the electronegativity, ion size, and ionic character of the HX molecule indicate their high chemical reactivity. Chloride ion is adsorbed on metal surfaces and replaces adsorbed oxygen or water molecules. This shifts the potential of the metal to more active (electronegative) values and causes breakdown of the passive state, mainly at elevated temperatures, which leads to the formation of pits. During attack of stainless steel (SS), chlorides of iron, nickel and chromium are formed; these are highly soluble in PA because of its complexing of cations of the transition group elements [14].

The values of the electrochemical parameters of the anodic polarization curve, such as i_{CC} which indicates corrosion activity, increase with increasing Cl^- concentration.

5.2. Influence of fluoride ion

The CaF_2 constituent of the fluoroapatite of PR reacts with H_2SO_4 during acidulation to produce HF which may form $(HF)_2$ and F^- ions, depending on the hydrogen ion activity of the solution according to the equilibrium:

$$\underset{\text{HF dimer}}{(HF)_2} \rightleftharpoons \underset{\text{Free HF}}{2HF} \rightleftharpoons \underset{\text{Hydrofluoride ion}}{HF_2^- + H^+} \rightleftharpoons \underset{\text{Fluoride ion}}{2F^-} + \underset{\text{Hydrogen ion}}{2H^+} \quad (9)$$

Dissolved aluminum compounds form acid-soluble aluminum fluoride complexes, for example, $(AlF_6)^{-3}$. Other reactions that decrease corrosion include formation of partially soluble metals fluoride complexes with Fe^{+3}, Mg^{+2}, Ca^{+2} and Na^+, and formation of insoluble fluoroaluminates and fluorosilicates that settle on metallic surfaces, such as Na_2SiF_6, $NaKSiF_6$, $MgNaAlF_6$, Na_3AlF_6 and $MgSiF_6$.

The main corrosion agent is the free fluoride ion, which is not complexed by cations. Reactive silica decreases the free HF by forming SiF_4 or H_2SiF_6. When the silica content of the PR is relatively less than the fluoride content, addition of reactive silica in the acidulation stage is recommended as a means of decreasing corrosion. This effect can be obtained also by mixing the rock with another PR that is high in reactive silica or aluminum compounds.

These effects have been demonstrated in laboratory corrosion test with 30% P_2O_5 (phosphoric acid) to which HF or complex fluorine compounds, such as H_2SiF_6 were added. The test included weight loss and electrochemical techniques. The electrode potential of UNS S31600 (0.16 V) becomes active in the presence of HF (−0.23 V), indicating a tendency to corrosion. The electrochemical parameters of the anodic polarization plot reveal anodic dissolution as a result of addition of HF and stable passive behavior in the presence of H_2SiF_6.

5.3. Influence of H_2S

Some PRs contain as much as 1% sulfide. In acidulation the sulfide is converted to hydrogen sulfide (H_2S), a weak acid that dissociates as in the equilibrium:

$$H_2S \rightleftharpoons H^+ + HS^- \rightleftharpoons 2H^+ + S^- \quad (10)$$

H_2S and the sulfide ion may act as reducing agents that affect the stability of the passive film on metals. Results of tests in a laboratory acidulation to which H_2S was added showed that the potential of S31600 fell slowly until it became negative, corresponding to active corrosion of the SS [15].

5.4. Influence of minerals: Erosion-corrosion

Erosion-corrosion (EC) is characterized in appearance by deep gullies, grooves, waves, rounded holes and valleys, exhibiting a directional pattern usually related to the direction of the fluid flow. These EC effects impair the protective film present on passive alloys, such as SS causing breakdown of passivity. On active alloys, such as CS, an active surface is maintained. In this way, the corrosion process is accelerated by mechanical removal of asperities, protuberances, and corrosion products, such as oxides or phosphates which otherwise might form a protective layer. A continuously renewed surface is generated, enhancing corrosion activity.

Corrosion, erosion and abrasion are frequent problems in chemical and mineral processing plants, leading to failures of equipment operating under severe hydrodynamic conditions. Many metals and alloys, such as CS-, SS- and Ni-base alloys, are susceptible to EC. All types of equipment that handle moving fluids, such as agitators, pumps, valves, nozzles, centrifuges, impellers and ball mills, are apt to undergo EC. The interaction between the chemical and the mechanical factors and their continuous effect on the metal surface causes EC [16].

6. Corrosion in phosphoric acid production

WPA plants operate under severe conditions that include elevated temperatures and rapid heat transfer, high acid concentration, agitation and circulation of liquids containing erosive suspended solids, aeration, formation of foam, and volatilization of corrosive acidic vapor that condenses on cooler metallic surfaces. Destructive corrosion results from combinations of these factors in which their combined action is greater than the sum of their separate actions.

PA is a mineral, nonoxidizing acid; its first dissociation follows the equilibrium:

$$H_3PO_4 \rightleftharpoons H^+ + H_2PO_4^- \quad Ka = 7.52 \times 10^{-3} @ 25°C \quad (11)$$

The corrosivity of pure PA solutions increases with increase in concentration to about 60% H_3PO_4 and then decreases with further increase in concentration. This behavior parallels the hydrogen ion concentration which increases with increase in acid concentration to about 50% H_3PO_4 and then decreases as the equilibrium in concentrated acid shifts to the left with further increase in acid concentration. H_2SO_4 solutions behave similarly. In both PA and H_2SO_4 there is a fairly good correlation between hydrogen ion concentration and the rate of corrosion.

In the WPA process, the filter acid (30% P_2O_5) is concentrated and converts into the PA product (52% P_2O_5). These acids are very corrosive, particularly to heat-exchanger surfaces where temperatures and velocities are high. During the concentration, part of the salts, such as fluoride compounds, precipitate and either from scale on the surfaces of the evaporation or remain in suspension. Fluorides are evolved with water vapor:

$$H_2SiF_{6(aq)} \xrightarrow{\text{heat and acid}} SiF_{4(g)} + 2HF_{(g)} \tag{12}$$

and Cl^- ions are removed as HCl vapor.

The corrosivity of halide ions in pure 30 and 50% H_3PO_4 decreases in the order $F^- > Cl^- > Br$ in accordance with the chemical activity of the halides. In 70 and 85% H_3PO_4 (52 and 62% P_2O_5), however, the corrosivity of the halides is in the order $Cl^- > F^- > Br$.

This result reflects the formation of stable monofluorophosphoric acid, which complexes F^- and decreases its corrosivity. The following equilibrium is established:

$$H_3PO_4 + HF \rightleftharpoons H_2PO_3F + H_2O \tag{13}$$

In 30 and 50% H_3PO_4, the fluorophosphate complexes hydrolyze and yield free active F^- which promotes corrosion.

A rise in temperature increases the rate of the chemical reactions that occur in the acidulation of the PR and the concentration of the acid. Higher temperatures in the reactor accelerate the corrosion of SS agitator and pumps, and higher temperatures in the concentration stage lead to increased corrosion in heat exchangers, pipes and pumps. To minimize corrosion, nonmetallic materials are used wherever possible; these include impregnated impervious graphite, carbon bricks, Teflon coatings and fiber-glass-filled phenolic resins.

Breakdown of the passive state may occur at elevated temperatures. For example, S31600 is resistant to 85% H_3PO_4 between 25 and 90°C, but at higher temperatures the passive layer is damaged and corrosion increases.

Agitation produces high velocity, turbulence, impingement, cavitation and erosion; all of these increase corrosion. The destructive effect that results from the combination of mechanical wear and electrochemical corrosion usually is greater than the sum of the two factors acting separately [17].

The abrasive solids, gypsum crystals, silica, and unreacted phosphate rock in the reactor slurry which contains about 35% solids, severely attack agitators, pump impellers and casings, and pipes, causing EC [18].

The ability of the SS- and Ni-based alloy passive films to protect the WPA plant equipment depends on their resistance to mechanical wear and their rate of regeneration when destroyed or damaged. The condition of the protective film depends on its interaction with the WPA constituents, such as Cl^-, F^- and oxidizing agents.

PR contains organic matter from fossil organisms, but its nature is not clearly defined. Its amount ranges from 0.5 to 2.3% organic carbon. Part of the organic matter may have been introduced as flotation agents that were adsorbed during beneficiation of the PR. Chemicals that control foaming, such as amines, fatty acids, and esters, are added in the acidulation step also contribute organic matter.

The effects of these different kinds of organic matter on corrosion in WPA have not been investigated thoroughly. They affect corrosion by forming films of organic matter on metallic surfaces and decreasing or increasing corrosion according to their chemical nature and physicochemical properties, such as adsorption.

6.1. Erosion-corrosion devices and measurements

In equipment handling, in particular in moving corrosive fluids containing suspended solid particles, such as in slurries, the main phenomenon is EC. A particular type of localized corrosion is caused by the synergetic action of mechanical erosion and electrochemical corrosion. Several environments and equipment are affected by FC in local plants: handling and processing slurries, for example, phosphate ore washing; wet process phosphoric acid (WPA) production; periclase hydration; and carnallite and sylvinite slurries, used in potash production.

In order to understand the causes and mechanisms of EC phenomena and to measure its magnitude, the chemical and mineral industries apply different laboratory and plant devices which simulate service conditions and sometimes increase the severity of the chemical, mechanical, and hydrodynamic factors involved in EC.

Many EC-measuring devices have been designed and constructed in the last years and are frequently reported in the literature [19]. They are based, in principle, on simulation of the mechanical forces acting on the plant equipment surface and a measurement of the mechanochemical effect on a static or dynamic metallic specimen exposed to a moving fluid The FC effect is assessed by physical changes (weight, length, volumes roughness), by an electrochemical measurement or by applying both techniques simultaneously or successively.

6.2. IMI erosion-corrosion measuring devices

During the past years the Corrosion Laboratory of IMI (TAMI)—Institute for Research and Development was engaged in the study of corrosion in several industrial plants handling aggressive slurries and in the selection of EC-resistant alloys to be used in these plants. Since EC is the main type of corrosion encountered in these systems, it was necessary to develop devices to study and measure FC phenomena in these slurries. Accordingly, two instruments were developed, built and applied in several plants.

6.2.1. Erosion-corrosion unit (ECU)

This instrument simulates EC phenomena occurring in equipment handling flowing fluids, enhanced by suspended solids in slurries or entrained solids in brines. A jet of slurry or brine is impinged continuously on the curved surface of the submerged rotating specimen, through a nozzle. The instrument measures corrosion potential and instantaneous corrosion current by the polarization resistance technique in accordance with the ASTM Practice for Conducting Potentiodynamic Polarization Resistance Measurements [20].

Figure 4 presents a schematic diagram of the apparatus, which consists of three parts: the cell, the agitated vessel and the electrochemical measuring instrument.

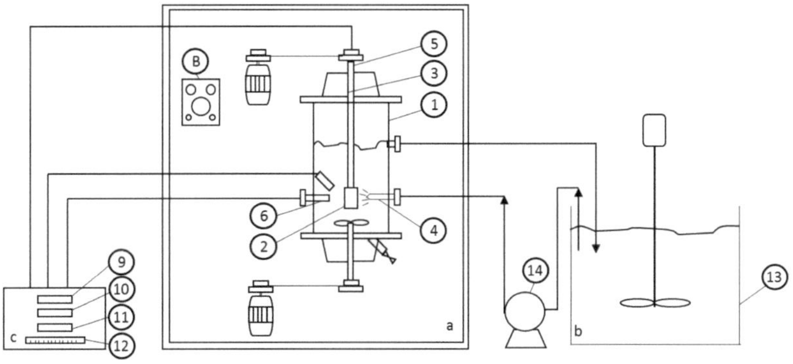

Figure 4. Diagram of erosion-corrosion unit/ECU.

The cell (a)—in a polypropylene vessel (1) (diameter 12 cm, height 18 cm) the alloy under test in the form of a cylinder (2) (diameter 20 mm, height 6 mm) is rotated about a vertical axis (3) while a jet of fluid is impinged continuously onto its curved surface through an interchangeable nozzle (4). The specimen and its mercury contact (5) are kept in place by means of a hollow rod holder. Immediately adjacent to the cylinder are an auxiliary platinum electrode (6) and a reference electrode (7).

The effects due to rotation, impingement and simultaneous rotation and impingement are determined by varying the specimen's rotational velocity and the jet's impact energy. Agitated vessel (b)—The cell is connected to an agitated slurry (B) which supplies the solids-containing fluid via a pump (14).

The electrochemical measuring instrument (c)—This instrument measures the potential (9) of the specimen and the corrosion current (10) which flows between the specimen (2) and the auxiliary electrode (6) when a fixed, 20-mV polarization potential (12) is applied to the specimen. The potential is set with respect to an Ag/AgCl or an SCE reference electrode (6) with a polypropylene body. The instrument also measures the oxidation-reduction potential E_{ORP} of the fluid. The electronic instrument was designed by Ch. Yarnitzky, Technion-Israel Institute of Technology, and built by Opal Co., Israel.

6.2.2. *Erosion-corrosion tester (ECT)*

The instrument applies rotational sliding abrasion on a static metallic specimen to simulate the corrosive wear effects of the solids-containing fluid on the surface of the equipment. The abrader material and its shape vary according to the solids type, the plant equipment and their interactions. The magnitude of the effect on electrochemical corrosion is measured by the polarization resistance technique to determine the instantaneous corrosion rate and by weight loss to assess the contribution of mechanical erosion.

The tester described in **Figure 5** consists of two parts:

The EC cell (a)—the cell (3) is a polypropylene cylinder, 40-cm high and 10 cm in diameter. It contains three electrodes; the working electrode(5) in the form of a disk, made from the alloy

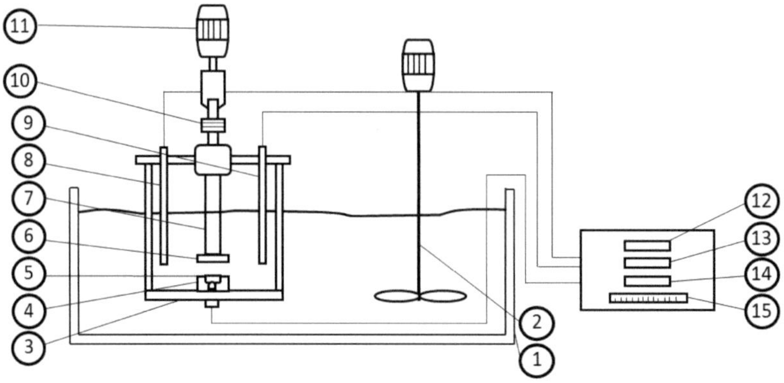

Figure 5. Diagram of erosion-corrosion tester/ECT.

under the test and the Ag/AgCl reference electrode (8) and auxiliary platinum electrode(9), both immediately adjacent to the disk (5). The cell is immersed into the vessel (1) containing the agitated fluid (2). A cross-shaped, rigid polypropylene abrader (6) or soft rubber wiper (6a) is attached to the base of the rotating shaft (7). The solids are swept outwards, and their particles are ground on the specimen surface. The shape of the abrader (and wiper) permits the particles to be swept onto the specimen surface. The severity of the mechanical action is controlled by weights (10) mounted on the rotating shaft and by increasing the rotational velocity of the shaft.

Electrochemical measuring instrument (b)—This instrument measures the open-circuit potential (12) of the specimen and the corrosion current (13) which flows between the specimen (5) and the auxiliary electrode (9) when a fixed, small polarization potential (15) is applied to the specimen, set with respect to the reference electrode (8).

6.3. Phosphoric acid production

Wet process phosphoric acid (WPA) is produced by attack of phosphate ore by concentrated sulfuric acid. Impurities, such as Cl^- and F^- increase the corrosivity of the medium. EC appears mainly in the reaction stage, where an aggressive slurry, containing 42% PA (H_2PO_4), 2% sulfuric acid (H_2SO_4) and 30% solid particles (gypsum, sand, quartz, and undissolved rock) is processed at 75°C (167°F). The main equipment undergoing EC are the reactor agitators and pump impellers, made of an austenitic stainless steel (Fe-20Cr-25Ni-4.5Mo-1.5Cu), in short, Fe-20Cr-25Ni.

The EC instruments were employed in WPA pilot plants to evaluate the corrosivity of various phosphate ores and for selection of EC-resistant alloys to be used for the fabrication of industrial plant equipment.

In an industrial plant trial, the ECT was applied to assess the effect of the addition of siliceous clays to the WPA reaction system, to complex the free hydrofluoric acid and minimize its corrosivity. AISI 316 SS high-alloyed austenitic stainless steels (Steel A, Fe-20Cr-25Ni-4.5Mo-1.5Cu, and steel B, Fe-17Cr-31Ni-3.5Mo-1.5Cu) and a Ni-base alloy (Ni-15Cr-16Mo-5Fe-2.5Co-4W) were tested. The results of several runs, with and without addition of clay, indicate clearly the significant reduction of EC rates in the presence of clay. The ECT was set at a rotation speed of 100 rpm and the load on the abrader was 1.1 kg/cm^3. The EC rate was measured by the linear polarization resistance technique.

Anodic polarization plots were obtained at a scan rate of 60 mV/min of Fe-20Cr-25Ni alloy, obtained both on a static specimen and on a specimen being abraded by the ECT. The current fluctuations, Plot b, indicate a rapid sequence of breakdown of the passive film by the mechanical action of the hard, sharp particles in the slurry, followed up by the formation of a protective oxide film. On the other hand, Plot a shows stable and smaller currents. Both plots reach same value in the transpassive region. A singular behavior was reported by Matsumura et al. during erosion-corrosion measurements.

7. Applications of phosphoric acid and phosphates

The great variety of PRs, industrial production process and plants of PAs and its distinct applications in critical reactors of the global economy, for example, fertilizers for the agricultural grouts, treatment of metallic machinery, addition to food and beverages, products for medicinal uses and for chemical treatment in environments and industries are depicted in **Figure 6**. A typical PR mine is presented in **Figure 7**.

- **Fertilizers**. WPA is the most important intermediate in the fertilizer industry since it is a major constituent of triple superphosphate, ammonium phosphate and mixed NPK fertilizers. Industrialized countries produce PK and NPK fertilizers for domestic utilization and for export; they are also supplied as slow-release fertilizers, as special fertilizer for fertigation (fertilization combined with irrigation) and for foliar application. Fertilizers may be acidic, neutral or basic; their pH and hygroscopicity affect their corrosiveness in the presence of moisture [21].

Ground PR is directly dispersed in agricultural fields in tropical regions to neutralize the soil's natural acidity.

- **Metallurgy**. Chemical conversion coatings are applied on steel and aluminum surfaces for protection against corrosion. Phosphating solutions, containing PA with special active additives are employed for protection of steel vehicles, office furniture, aircraft, merchant and military ships and machinery. The phosphate coating ensures the adhesion and performance of posterior painting [22, 23].

Phosphate conversion coating is the most widely employed technique to afford good corrosion protection to sheet steel, in particular in the automotive industry. PA is an important industrial acid utilized in many industries. In metallurgical applications, it is used for pickling and supplementary treatment of steel sheeting used for steel car bodies and steel machinery. The phosphate layer consists of numerous crystals of different sizes, which implies the presence of voids between these crystals. Porosity is usually quite low: that is, 0.5–1.5% of phosphate coating.

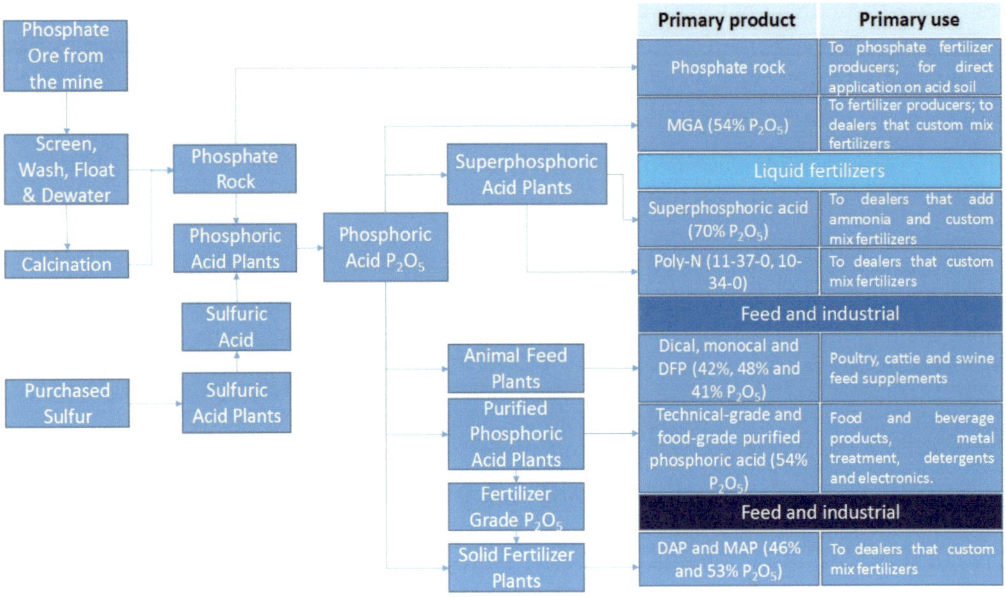

Figure 6. Phosphoric acid process, products and uses.

Figure 7. Phosphate rock mine.

Solutions of PA are employed for removal of rust from corroded surfaces; the black phosphate coating improves their corrosion resistance. PA solutions are utilized for cleaning and sanitation of equipment and machinery in industrial plants, for example, dairy plants; for electropolishing of SS, aluminum and copper alloys; and as a flux component in soldering. Fuel cells for electricity generation operate with PA as a liquid electrolyte.

- **Chemicals**. PA is used for many chemical devices and operations: to remove mineral deposits, to clean hard water stains, to etch solution nitride in micro-fabrication, in hydroponics to lower the pH of nutrient solution, as an etching agent for semiconductors, in cosmetic and skin-care products to adjust the pH, for drinking water treatment, for preparation of synthetic rubber, for leather tanning, as an additive for varnishes, pigments and paints. It is used as an additive in the manufacture of fire bricks, fire retarding agents, ceramic colors, and as a catalyst in polypropylene polymerization and to enhance the setting action of synthetic resins. Laundry detergents contain soluble phosphates. Phosphonates, organophosphorus compounds from PA, are used as corrosion and scale inhibitors for water treatment and in the desalination industry.

- **Food and beverages**. Pure food grade PA [24, 25] is an additive to food sauces, mayonnaise and fruit juices, and is used to acidulate cola-type beverages. Calcium phosphate salts are added to baked goods and white, soft cheeses to avoid water segregation [26]. It is used in sugar and edible oil refining and for bacteria control in food processing. Sodium and potassium phosphate salts serve as food preservatives.
- **Medicine**. PA combined with zinc powder forms zinc phosphate, which is used as a dental cement. In orthodontics it is applied to clean and roughen the teeth before inserting brackets and other dental devices. Eliminating plaques and whitening the teeth are done with PA derivatives. Orthopedic metallic implants are covered with calcium phosphate to promote their integration with osseous tissue. Phosphatic cements are applied in surgery of bone systems. Also, phosphate salts are used to reduce pain in sensitive teeth.

Author details

Benjamín Valdez Salas, Michael Schorr Wiener* and Juan Ricardo Salinas Martinez

*Address all correspondence to: mschorr2000@yahoo.com

Institute of Engineering, Universidad Autónoma de Baja California, Mexicali, Baja California, México

References

[1] Dillon CP. Corrosion Control in the Chemical Process Industries. 2nd ed. Houston, TX: NACE; 1993

[2] Lehman U. Environmentally sustainable recycling of acid for pickling of stainless steel. ThyssenKrupp Techforum. 2005;**1**:26

[3] Gilmour R. Phosphoric Acid: Purification, Uses, Technology, and Economics. Boca Raton, FL: CRC Press; 2013

[4] Ross RW. Corrosion by phosphoric acid. In: ASM Handbook. Stephen D. Cramer and Bernard S. Covino, Corrosion: Environments and Industries. Vol. 13C. Materials Park, OH: ASM International; 2006

[5] Schorr M. Corrosion control in WPA production using Phalaborwa igneous phosphate rocks. Phosphorous & Potassium. 1993;**184**:23–33

[6] Schorr M, Valdez B. The phosphoric acid industry: Equipment, materials, and corrosion. Corrosion Reviews. 2016;**34**:85–102

[7] Jasinski SM. Phosphate rock. USGS Minerals Information. Personal Communication; February 2009

[8] Charrach J, Schorr M, Weintraub E. Corrosion and scaling behavior in Dead Sea basin saline waters. Corrosion Reviews. 1990;**9**:293–352

[9] Baniel A, Blumberg R, Alon A. Process for preparation of substantially pure phosphoric acid. US Patent: 1965. No. 3 338 674

[10] Baniel A. HCL Clean Tech. Jerusalem, Israel. Personal Communication; January 2015

[11] Blumberg R. Liquid-Liquid Extraction. London: Academic Press; 1998

[12] Schorr M, Valdez B, Zlatev R, Stoytcheva M. Erosion-corrosion in phosphoric acid production. Materials Performance. 2010;**49**:56–59

[13] Schorr M. Corrosion Manual, Corrosion Testing and Control in Fertilizer Intermediates and Products. Part One: Wet Process Phosphoric Acid. USA: International Fertilizer Development Center; 1981. pp. 1–98

[14] Shi W, Xiang S, Li YL, Yang M, Hu YN, Wang QD. Effect of temperature on electrochemical behavior of stainless steel in phosphoric acid. Applied Mechanics and Materials. 2013;**351**:1072–1076

[15] Bellaouchou A, Guenbour A, Benbachir A. Corrosion behavior of stainless steel in phosphoric acid polluted by sulfide ions. Corrosion. 1993;**49**:656–662

[16] Schorr M, Valdez B, Zlatev R, Stocycheva M. Phosphate ore processing for phosphoric acid production; classical and novel technology. Mineral Processing and Extractive Metallurgy. 2010;**119**:125–129

[17] Schorr M, Valdez B, Zlatev R, Santillan N. Agitator corrosion in wet phosphoric acid production. Materials Performance. 2007;**46**:50

[18] Hael AE, Mohammad AS, Hassan H. The effect of motion on the behavior of corrosion stainless steels in industrial phosphoric acid. Open Access Scientific Report. 2012;**1**:9

[19] Jang W, Pozzo RL, Iwasaki I. Technical note: Estimation of corrosive wear of grinding media by electrochemical measurements. Corrosion. 1988;**44**(11):836–838

[20] ASTM Standard G59-78. Standard Test Method for Conducting Potentiodynamic Polarization Resistance Measurements. West Conshohocken, PA: ASTM International; 1984

[21] Agarwal DC. Phosphoric acid production for fertilizer applications. Stainless Steel World. 2002;**14**:58–65

[22] Sanchez-Tovar R, Montanes MT, Garcia-Anton J, Guenbour A. Galvanic corrosion of the base AISI 316l/micro-plasma arc welded AISI 316l in polluted phosphoric acid media at different temperatures. International Journal of Electrochemical Science. 2011;**6**:5550–5564

[23] Santana I, Pepe A, Jimenez-Pique E, Pellice S, Ceré S. Silica-based hybrid coatings for corrosion protection of carbon steel. Part I: Effect of pretreatment with phosphoric acid. Surface and Coatings Technology. 2013;**236**:476–484

[24] Rotem Amfert Negev. Food grade phosphoric acid 85%. Product specification. Rotem Amfert Negev; 2012

[25] Li YL, Xiang S, Zeng HT, Wang JP, Wang QD. The corrosion behavior of 304L and 316L stainless steels in food grade phosphoric acid solutions. Applied Mechanics and Materials. 2012;**109**:28–31

[26] Calcium Phosphates [Internet]. 1997–2016. Available from: http://www.chemeurope.com/en/encyclopedia/Calcium_phosphate.html [Accessed: February 20, 2015]

Chapter 5

Occupational, Public and Environmental Radiological Impact Caused by the Phosphoric Acid Industry: The Case of Huelva (Spain)

José Luis Guerrero-Márquez,
Fernando Mosqueda Peña, Juan Mantero,
Guillermo Manjón, Rafael García-Tenorio and
Juan Pedro Bolívar

Additional information is available at the end of the chapter

http://dx.doi.org/10.5772/intechopen.68567

Abstract

The production of phosphate fertilizers usually uses as raw material sedimentary phosphate rock, which contains enhanced concentrations from U-series radionuclides about 10–100 times higher than unperturbed soils. This fact implies the need for evaluating the radiological implications of this activity. In our case, the study has been performed in a large fertilizer industrial complex located at Huelva town (SW of Spain), where sedimentary phosphate rock has been processed since 1965 to 2010, generating annually an average of about 2.5 million tons of a by-product called phosphogypsum (PG), which has been stored in big stacks 1 km away from Huelva city, covering 1000 ha. The fluxes of the radionuclides of interest along the production process and the effective doses received by the workers have been determined. In addition, the radioecological impact associated to the waste management strategy followed has been evaluated.

Keywords: NORM, radiation doses, phosphogypsum, radioecological impact

1. Introduction

The production of phosphoric acid (PA) from phosphate ore deposits is essential because phosphorus is an essential element to all living systems, and especially for the plants. The phosphate fertilizers and animal feeds (among others) are manufactured by using the PA as raw material. The bioavailable phosphorus is usually present in very low quantity in most agricultural soils, and for this reason, it is necessary to add it as soluble phosphate.

Total global phosphate reserves are estimated to be around 1.6× 10^9 tons. Most of these deposits are from sedimentary origin, with as little as 4% being of igneous origin [1]. Phosphatic materials are usually characterized in terms of their P_2O_5 concentration in percentage. This is done as a convenience and not as a strict chemical description. Phosphate ores generally have P_2O_5 concentrations ranging from 4 to 40%.

In simplified form, phosphate ore is converted into commercial products using the following three main process steps:

i) After being mined, phosphate ore is beneficiated to produce a concentrate known as phosphate rock (PR).

ii) According to International Fertilizer Industry Association [2], as much as 85% of the phosphate rock produced in step (i) is converted into intermediate or final products using a process of acid digestion known as the "wet process." A relatively small amount of PR is converted directly into elemental phosphorus by reduction in an electric arc furnace in a process known as the"thermal process," in order to obtain a more pure PA. It is estimated that 71% of all PR produced is processed into phosphoric acid, with the generation of a by-product called "phosphogypsum" (PG). In addition, the 24% of PR is processed directly into fertilizer without the generation of PG, whereas the remaining 5% is converted directly into various other products. The annual production of P_2O_5 in the form of PA is more than 30 million tons.

iii) Most of the phosphoric acid produced in step (ii) (75–90%) is subjected to further chemical processing to convert it into fertilizer. The fertilizer products so derived from PA thus account for some 55–60% of total phosphate rock production. Of the remaining 10–25% of PA produced in step (ii), about one-half is processed into animal feed supplements and the other half into a variety of other products.

The production of PA for the manufacture of phosphate fertilizers usually use sedimentary phosphate rock as raw material, which contains generally enhanced concentrations from U-series radionuclides; about two orders of magnitude in relation to unperturbed soils. This implies huge fluxes of radionuclides along the production process, and the generation of wastes and by-products enriched in natural radionuclides. For that reason, the evaluation of the occupational, public and environmental radiological implications of this industrial activity is needed.

This evaluation has been performed for the authors in a large fertilizer chemical industrial complex, located at Huelva town (SW of Spain), where sedimentary phosphate rock has been processed since 1965 to December 31, 2010, generating annually about 2.5 million tons of phosphogypsum waste (PG), which has been stored in big stacks (around 1000 ha and 100 million tons of PG are stored at this repository). The fluxes of the radionuclides of interest along the production process and the effective doses received by the workers have been determined. In addition, the radioecological environmental impact associated to the waste management strategy has been also evaluated.

2. Phosphoric acid production at Huelva

2.1. Description of the PA production

At Huelva five phosphoric acid plants were working, consuming annually around 1.5 million tons of phosphate rock, and producing 0.5 million tons of PA (expressed as P_2O_5), and 2.5 million tons of PG. PG is an unavoidable by-product in the production of PA by the sulphuric "wet-acid" method, where 5.5 tons of PG per ton of P_2O_5 are produced.

The process for obtaining the phosphoric acid from the phosphate rock can be described schematically by the following chemical reaction [3]:

$$Ca_3\,(PO_4)_2 + 3H_2\,SO_4 \rightarrow 2H_3\,PO_4 + 3CaSO_4 + 48.9\ \text{kcal} \tag{1}$$

In the Huelva phosphoric acid plants the fluorapatite was mainly used as raw material, which in the process conditions induce the production of the phosphogypsum in the di-hydrate form, according to the following general reaction [3]:

$$Ca_{10}\,(PO_4)_6\,F_2\,CO_3\,Ca + 11H_2\,SO_4 + 21H_2\,O \rightarrow 6H_3\,PO_4 + 11CaSO_4{\cdot}2H_2\,O + 2FH + CO_2 \tag{2}$$

The industrial process can be divided into four different steps (**Figure 1**): (1) grinding and conditioning of the phosphate rock, (2) dissolution of the phosphate rock with diluted sulphuric acid (60%), (3) filtration, or PG separation from phosphoric acid (liquid fraction), and (4) washing of PG by water.

The grinded PR is introduced into the digesters where the phosphate rock is attacked (dissolved), and a pulp is obtained containing mainly PA in solution, and solid PG. This pulp is filtered and PA (24% P_2O_5) and PG are separately obtained. Then, three successive backwashes are done in order to extract the remaining PA still contained in the phosphogypsum. The first washing is carried out with PA at 12% P_2O_5 where phosphoric acid with 27% P_2O_5 is obtained. The second washing is done with 5% P_2O_5 to produce 12% P_2O_5 which is used in the first washing. Thirdly,

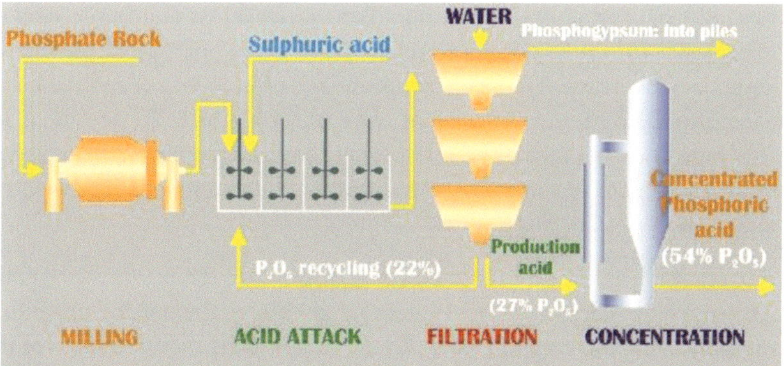

Figure 1. Industrial process of PA production.

a final washing is performed with hot water to produce 5% P_2O_5, which is used in the second washing.

Finally, the 27% P_2O_5 (production acid) is concentrated by evaporation for obtaining commercial acid (54% P_2O_5), since it is the normal concentration needed for the fabrication of the most demanded phosphate fertilizers: monoamonium phosphate (MAP), diamonium phosphate (DAP), and complex fertilizers (NPK).

2.2. Description of PG piles

Although during the last years research and attempts have been done for valorization of the PG, the use of the generated one in the Huelva factories has been until now quite limited and mostly in agriculture as a soil amendment [4]. Consequently, due to the lack of any market for this compound, the great majority of the generated PG in Huelva has historically been considered as a waste with the following management policy: (a) During the first 30 years of production (until December 1997), 20% of the PG generated was released directly into the Odiel Channel of the Huelva estuary, whereas the remaining 80% was transported from the phosphoric acid production plants in suspension with sea water to a neighboring marshland area (2 km away from the plants and located in the confluence of the mouths of the Odiel and Tinto rivers), where it was disposed of by means of decantation with the transporting waters draining directly into the estuary [5, 6]; (b) since beginning 1998, all the PG produced in Huelva has been stored in the PG piles following the OSPAR convention [7, 8]. The PG was transported to the disposal area with fresh water, which was reintroduced in the industrial process following a closed cycle, in order to prevent any impact on the estuarine ecosystem [9].

Since December 31, 2010, the production plants of phosphoric acid have been closed, and, as a consequence, the production of PG has been halted. The legacy from these plants involves a major area where millions of tons of PG have been accumulated, and can be considered as formed by four different areas called Zone 1, Zone 2, Zone 3 and Zone 4, as it is shown in **Figure 2**.

Zone 1 is a 450 ha sector and it was the first area submitted to regeneration work. It is calculated that about 12×10^6 tons of PG, including unknown amounts of urban wastes and mining wastes, were released in this sector. Restoration tasks were performed during 1990 by the Andalusia Regional Government in order to minimize both the visual and the environmental impact associated to the different wastes that were released there. To do that, a soil cover of about 30 cm average thickness and a vegetal coverage were added above the naked PG surface. Furthermore, several tenths of soil hills (1.5 m high) were scattered above the soil cover layer in order to allow small tree species to grow.

In Zone 2 (270 ha), the wastes were released until 2010 inside the water recirculation system. In this zone, the gypsum stacks reached 15 m in height above the natural ground, and it continued to grow following a pyramid shape. As the height increases, the sides of the pyramid are covered with a soil layer and a vegetal coverage, in order to minimize the visual impact.

Zone 3 (180 ha) is close to Zone 2. It is only conformed by PG, without any soil coverage.

Figure 2. Aerial photograph of the area of the Tinto river marshes. (1) Settling ponds, dumping area of FY; (2) regulating reservoir, (3) perimeter channel, and (4) pumping station.

Zone 4 was submitted to a similar restoration process than Zone 1. More detailed, different kinds of industrial and urban wastes were added above the PG: in the first place, a building debris layer was added; thereafter, different kinds of industrial wastes were added and finally, a soil layer like that of zone 1 for different vegetable species can grow on this cover. The total thickness of the coverage (including the three layers: wastes, debris, and soil) is more than 1 m.

Furthermore, one of the sectors contained within the PG stacks was contaminated some years ago by the radioactive ashes produced during the Acerinox Steel factory accident in 1998. An industrial ^{137}Cs radioactive source was accidentally melted with iron wastes in the steel factory. These wastes were treated in a purifier plant in the industrial complex of Huelva. Finally, the wastes and a certain amount of their ashes were released in Zone 4 and mixed with the PG. As a consequence of the releases, the Spanish National Wastes Enterprise (ENRESA) sealed the sector and restoration works were developed consisting in a cover with a layer of clay to minimize permeability and migration of radioactive cesium. Finally, different vegetal species were planted in the sector, and a vigilance program was established.

Besides, from Zone 3, it has been possible to observe liquid leakages (filtrations), mostly with a certain flow, into the Tinto River and adjacent marshlands, while from Zone 2, these leakages are clearly minor in number and diffused after overflowing the perimeter channel barrier.

Currently, there is an environmental plan under study in order to restore Zones 2 and 3; for the total application of which, a duration of 10 years has been estimated. It is expected that the planned restoration should solve the environmental impact on the neighboring areas.

2.3. Samplings

2.3.1. Phosphoric acid plants

In relation to the studies made inside the fertilizer plants, several sampling campaigns have been developed in the phosphoric acid factory located at Huelva (Spain), with the aim of analyzing the distribution [10–12] and levels of radionuclides in the materials involved in its production and evaluating the occupational, public and environmental radiological impact of this factory. Solid, liquid and air-filter samples were collected in: (a) the four steps of the PA process: (1) milling and grinding, (2) acid attack or digestion, (3) filtration and washing, and (4) PA concentration; and in (b) the plants devoted to the production of monoamonium phosphate (MAP) and diamonium phosphate (DAP) fertilizers. The solid samples included several matrices, from the phosphate rock until commercial phosphoric acid (54% P_2O_5).

The water samples were collected at the regulating reservoir associated to the big active pile where the PG was finally stored until 2010. This water is used for the phosphogypsum pumping into the ponds, and it was collected over a year at a rate of one sample per 2 months, in order to evaluate the possible temporal evolution in its radionuclide contents.

In order to estimate the committed effective doses received by the workers of the PA, MAP and DAP plants due to inhalation, aerosol filters were collected from different representative places of the plants. The aerosol collections were performed with Andersen PM10 high-volume samplers. The external radiological effective doses were measured by using a radiation monitor Universal Monitor (UMo) model LB 123. The "Universal Monitor" (UMo), manufactured by Berthold, is a monitor designed specifically for low dose rates. It is equipped with a probe which works on proportional mode and allows the measurement of both dose rate and integrated dose.

2.3.2. Phosphogypsum stacks

The research groups involved in the writing of this chapter have been performing research and monitoring studies associated to the PG stacks and their neighboring environmental compartments since the end of the 1980s.

Several sampling campaigns have been done inside the PG stacks in order to characterize the material accumulated, to analyze the spatial degree of variation in the levels of the radionuclides associated [4, 13], and to evaluate its radiological impact through external radiation, inhalation of particle matter and radon emanation. And also a detailed sampling campaign was done after the halting of the PG production in order to evaluate the radioactive content of the liquid leakages of the piles and the associated formed efflorescences [14].

In addition, dozens of sampling campaigns have been performed in the PG neighboring environments as are summarized in the following paragraphs. In this sense, more than 10 sampling campaigns have been performed from the end of the 1980s until now in the estuaries of the Odiel and Tinto rivers to evaluate the temporal evolution of the environmental radioactive contamination generated by the PG piles and the effluents from fertilizer plants. The majority of these sampling have involved the collection of waters (dissolved and particulate matter fractions), and superficial sediments, although special sampling campaigns were performed, for example, for the collection of several sediment cores in the estuaries to evaluate its possible use as historical archives of the radioactive contamination in the area [15]. In addition, sediments were collected in order to perform speciation studies [16]. In all the cases, several actions were taken in order to preserve the integrity and quality of the samples for radionuclide determinations (acidification of the waters after filtration, drying at room temperature of the sediments in order to avoid changes in the forms of association of the radionuclides to their different phases, etc.).

On the other hand, in addition to the studies carried out in the estuaries mostly with surveillance purposes, other sampling campaigns were performed in order to carry out more pure radioecological studies. In this sense, several sampling campaigns were performed for the collection of vegetation in the marshes surrounding the PG piles (halophytes plants) and the determination of natural radionuclide transfer factors (TFs) [17–19], and the collection and analysis of other biota samples (shellfish, shrimps, etc.).

2.4. Measuring techniques

The radioactive characterization of the great majority of the samples collected either in the production plants, in the PG stacks or in the neighboring environmental compartments (PG, waters, sediments, soils, efflorescences, vegetation, biota, etc.) have been performed by applying two independent techniques: both gamma-ray and alpha-particle spectrometry.

Gamma measurements in solid samples (PG, soils, sediments, biota, etc.) were carried out using along the time several gamma spectrometry systems, most of them based in High-Purity Germanium (HPGe) detectors, although in some specific cases when the amount of material was limited a well-Ge detector was used. The calibration of the gamma systems were performed following procedures as the described in Ref. [20], applying when was needed corrections due to self-absorption effects.

In the same solid samples, as well as in waters, Th-isotopes, U-isotopes and ^{210}Po activity concentrations were determined by alpha-particle spectrometry after the application of validated radiochemical methods, such as the described in Ref. [21], for the sequential isolation of the elements of interest. Alpha-spectrometric systems formed by independent chambers working in parallel and equipped with 450 mm^2 PIPS detectors were used for the measurements. After U and Th were isolated from the matrix under analysis, they were independently adapted for measurement by electrodepositing them onto stainless steel discs by applying the method of Hallstadius [22], whereas Po was self-deposited onto silver discs of 1 inch diameter using Flynn's (1968) method.

Along the 35 years of measurements in the area, other techniques have been also used for specific radionuclide determinations: U and Th elemental determinations, for example, have been performed in waters, PG and sediment samples by ICP-MS. ^{226}Ra and ^{210}Pb determinations in waters have been performed after radiochemical isolation by using the Liquid Scintillation Counting (LSC) technique or low-background gas-flow proportional counters; ^{222}Rn determinations have been carried out on the PG stacks using Rn monitors, etc.

All the procedures used for radionuclide determination have been periodically validated through the regular participation in intercomparison exercises organized at national (Spanish Security Council) and international (IAEA) level. Additionally, different quality control (QC) procedures were routinely applied: performance of replicate analysis, measurement of blank samples, measurement of certified reference materials, regular control of system's background, etc.

2.5. Radioactive characterization

2.5.1. Phosphoric acid plants

The sedimentary phosphate rock treated for the production of phosphoric acid is clearly enriched in radionuclides from the uranium series. They are at least 50 times higher than in representative unperturbed soils worldwide [23]. In addition, in the PR all members of the uranium series are in secular equilibrium (**Table 1**), as expected due to the fact that the material does not experiment any radionuclide enrichment during the physical processes applied before its digestion. On the other hand, the activity concentrations of the radionuclides belonging to the ^{232}Th chain in the treated phosphate rock are quite reduced.

During the digestion step, the first clear evidence of different behavior that the analyzed radionuclides from the U series show in the process is observed. A clear fractionation of the radionuclides contained in the mineral is produced between the liquid and solid fractions from the pulp samples collected from the digesters. The liquid fractions, corresponding mainly to the phosphoric acid formed at this step, are clearly enriched in uranium, whereas the concentrations of ^{230}Th, ^{226}Ra, and ^{210}Pb are very much lower or even negligible (this is the case of ^{226}Ra). On the contrary, solid samples contain the major proportion of the ^{230}Th, ^{226}Ra, and ^{210}Pb originally present in the phosphate rock, whereas the uranium isotopes concentrations are slightly lower. This solid fraction is mostly composed of the PG formed in reaction to the phosphate rock with sulphuric acid. However, it may also contain a proportion of non-attacked phosphate rock and some phosphoric acid occluded between the PG grains and not separated from the liquid samples.

The results obtained in the analysis of the samples collected in the filtration step allowed the confirmation of the different behavior of the uranium isotopes and their daughters previously drawn. By observing the data shown in **Table 1**, we can affirm that most of the uranium tends to be associated to the phosphoric acid fraction, whereas the majority of ^{230}Th, ^{226}Ra, and ^{210}Pb tends to be associated to PG. These conclusions are evident by observing the obtained activity concentrations measured in the phosphoric acid in relation to the P_2O_5 concentration.

The U concentration decreases in the PG samples according to their successive washings, which indicates that the remaining P_2O_5 of the PG is sequentially removed from the different

	^{238}U	^{234}U	^{230}Th	^{226}Ra	^{210}Pb	^{232}Th	^{228}Th	^{228}Ra
PR	1653 ± 21	1405 ± 504	1600 ± 87	1565 ± 124	1578 ± 95	26 ± 5	20 ± 3	21 ± 1
22% P_2O_5	870 ± 50	870 ± 50	95 ± 4	3.0 ± 0.4	74 ± 14	53 ± 11	1.7 ± 0.4	<1.8
Cake dig. L	1213 ± 186	1197 ± 189	305 ± 104	5 ± 1	73 ± 11	4 ± 2	5 ± 2	4 ± 4
Cake dig. S	490 ± 61	490 ± 61	650 ±125	607 ± 146	660 ± 185	15 ± 1	11 ± 2	12 ± 3
Pulp cake Solid	1340 ± 90	1330 ± 90	425 ± 78	4.0± 0.4	60 ± 12	6.5 ± 1.4	4 ± 0.5	<1. 0
Pulp cake liquid	150 ± 30	450 ± 30	750 ± 30	590 ± 25	590 ± 120	16 ± 2	11 ± 2	9. 4 ± 1.1
27% P_2O_5	1070 ± 110	1090 ± 110	230 ± 30	5,6 ± 0.5	78 ± 15	9.8 ± 1.8	3.9 ± 0.5	<1.8
22% P_2O_5	870 ± 50	870 ± 50	95 ± 4	3.0 ± 0.4	74 ± 14	53 ± 11	1.7 ± 0.4	<1.8
12% P_2O_5	470 ± 40	470 ± 40	80 ± 10	6,1 ± 0.5	<225	6.0 ± 1.5	<1.7	<1.9
5% P_2O_5	290 ± 20	290 ± 20	4,2 ± 0.3	1,8 ± 0.3	<135	0,17± 0.06	<1.0	<1.5
Gypsum	425 ± 152	439 ± 143	775 ± 47	635 ± 39	620 ± 16	26 ± 8	10 ± 3	10 ± 1
Sludges L	1675 ± 587	1670±608	750 ± 707	6 ± 2	65 ± 27	14 ± 12	11 ± 7	1.7 ± 0.1
Sludges S	1070 ± 226	1065±233	1455±771	2805±2114	1660±339	20 ± 11	21 ± 15	41 ± 33
32% P_2O_5	1300 ± 70	1320 ± 70	NM	3.6 ± 0.3	<135	NM	6.5 ± 0.4	<0.9
54% P_2O_5	1830 ± 180	1810 ± 180	NM	7.9 ± 0.5	109 ± 14	NM	13.2 ± 1.4	<1.0

Note: Uncertainty = standard deviation, L = liquid fraction, S = solid fraction, NM = not measured.

Table 1. Activity concentrations (mean values in Bq kg^{-1}) obtained for the samples collected along the milling, digestion, filtration, washing, and acid concentration steps.

washings. On the contrary, the concentrations of ^{230}Th, ^{226}Ra, and ^{210}Pb remain practically constant along the successive washings, showing this fact that this type of radionuclides are very hard bound on the PG fraction.

A clear linear correlation between both U and P_2O_5 concentrations was obtained by using the different materials involved in the phosphate fertilizer production process, with a slope of 35 ± 2 Bq kg^{-1} ^{238}U/% P_2O_5. This result demonstrate that the uranium follows the same route that the PA along the full fertilizer production process.

Another conclusion of these studies was that a high fraction of the uranium series radionuclides (mainly U and ^{230}Th) input in the P_2O_5 manufacturing arrives finally into the environment by the use of phosphate fertilizers in the agriculture.

2.5.2. Phosphogypsum stacks

The activity concentrations measured in the materials stored in the PG piles reflect the radionuclides fractionation in the PA manufacturing, being found for PG that the levels are in the sequence: ^{226}Ra ≈ ^{210}Pb ≈ ^{210}Po > ^{230}Th > ^{234}U = ^{238}U. Activity concentrations of 500–700 Bq kg^{-1} for both ^{226}Ra and ^{210}Pb, and 80–200 Bq kg^{-1} for ^{234}U and ^{238}U have been determined along the PG stacks [9]. However, the activity ratios found in the PG stacks are something different than the determined ones in the fresh PG obtained immediately after its formation, reflecting this facto the different behavior of the radionuclides analyzed along the time in the deposits. In this sense, the PG stored in the stacks is depleted in U in comparison with fresh PG, which indicates that U is weekly bounded to the PG, incorporating it either to the waters used for the transport of the PG from the factories to the piles, or to the waters interacting with the piles (rain waters, tidal waters from the estuary, etc.). On the contrary, the ^{226}Ra levels in the PG stacks are similar to the found ones in fresh PG, in correspondence with its low solubility.

Supporting the previous comments, to point out that the average activity concentration of ^{210}Pb and ^{226}Ra in the PG are similar in both Zones 2 and 3 of the PG stacks (**Figure 1**), while the average concentration of ^{238}U in PG of Zone 3 (95 ± 15 Bq kg^{-1}) is lower than that in the Zone 2 (220 ± 85 Bq kg^{-1}). This fact is due to the PG from Zone 3 has over time, lost a fraction of U higher than Zone 2 (the PG in zone 3 was stored earlier).

The high solubility of the U contained in the fresh PG is reflected in the extremely high concentrations of this element in the leaching waters from zone 2. High activity concentrations, although in minor extent than for U, were also found in these waters for other uranium-series radionuclides such as ^{210}Po and ^{230}Th. The levels of uranium in these leaching waters are 4 orders of magnitude higher than those obtained in the sea water [24], and even reach the value of 300 Bq L^{-1} for ^{238}U, while the levels found for ^{210}Po and ^{230}Th were 3 and 2 orders higher, respectively, than in sea water. These waters present extremely low pH (pH < 2), which makes it clear that, in the planned process of restoration, these internal PG waters should be carefully treated and neutralized before their release into the estuary. The ^{230}Th/^{238}U, ^{210}Po/^{238}U, and ^{210}Po/^{230}Th activity ratios in these leaching waters are clearly at lower levels than in the PG, reflecting the different way of bounding these radionuclides to the PG and their different solubility in acidic waters.

The uranium activity concentrations determined in leakage waters percolating from Zone 3 to the neighboring salt marshes are also high in comparison with the activity concentrations usually found in seawater, but remain lower than those found in the waters from the perimeter channel (at least three to five times lower), a fact that can be associated with the possibility of a mixture of contaminated waters coming from the PG pile with waters from the estuary of markedly lower activity concentrations. The leaking waters from Zone 3 are also characterized by their very high concentrations of radionuclides from the uranium series analyzed although the relative abundance of these radionuclides in them is different than those in the waters leaking from Zone 2, because since its formation (more than 15 years ago), uranium series radionuclides have been leaking from this zone, and today only a fraction of the radionuclides remains out of those originally present in the pile. This fraction is lower for the radionuclides less bounded to the PG structure, i.e., those more easily dissolved. This fact explains why the U-isotope activity concentrations found in these waters, being very high and relatively uniform, are at the same time one order of magnitude lower than the concentrations determined in the perimeter channel of Zone 2, and lower than the highest values found in some waters percolating into the surroundings of Zone 2.

On the other hand, a big range of activity concentrations were found for different radionuclides of the uranium series in efflorescences collected in the surroundings of Zone 3 and 2 of the PG piles was found, with U-isotope activity concentrations up to 8.7 kBq kg^{-1}, ^{226}Ra activity concentrations up to 2.2 kBq kg^{-1}, and ^{210}Pb activity concentration up to 70 kBq kg^{-1} [14]. These results in efflorescences can be explained by taking into account the fact that they are formed by the evaporation of the diffused waters escaping from the piles that, in the situation of near dryness, induce the precipitation of a big proportion of the associated metals and radionuclides. The formation mechanism of these efflorescences explains in fact why the activity concentrations of ^{210}Pb are higher than the ^{238}U and ^{226}Ra activity concentrations in most of the samples analyzed, as is reflected in the $^{210}Pb/^{238}U$ and $^{210}Pb/^{226}Ra$ activity ratios which are generally higher than 1, because it is well known that the precipitation trend of Pb is different than that for Ra and U [25].

The activity concentrations of ^{226}Ra in the great majority of efflorescence samples analyzed are clearly lower than the activity concentrations of ^{238}U, ^{210}Pb, and ^{210}Po, and only in a very limited number of cases exceed the average concentrations found in the PG disposed of in both the analyzed zones, allowing the rejection of the hypothesis that the efflorescences in the surroundings of Zone 2 and 3 are formed by particles of PG with the smallest grain size transported by the leakage waters; it is well documented that the ^{226}Ra activity concentrations in the PG finest fractions are clearly higher than the found ones in the bulk PG.

Finally, it is worth to mention that the impact of the PG stacks in the atmospheric compartment has been also evaluated through radionuclide determinations performed in aerosol samples collected over the piles and through ^{222}Rn determinations. The obtained results indicate small enhanced concentrations from the uranium series radionuclides in the aerosols collected over the piles, although this fact is not reflected in significant either occupational or public doses, due to the very low aerosol concentrations generated in the area. The main reason is that the PG produces a hard crust on the surface of the piles that avoid the resuspension of material.

On the other hand, the concentration levels of ^{222}Rn over the piles are not statistically different to the values found in background reference areas, indicating that the generated ^{222}Rn in the piles as a daughter element of ^{226}Ra, when escapes from the piles, quickly experiments its dilution in the atmosphere, process that is favored by the breezes that usually affected the coastal area where the piles are disposed. A high spatial variability for ^{222}Rn exhalation was found in the PG piles, although the average values are one order of magnitude higher than the found ones in neighboring background areas [26]. This result confirms the importance of the commented dilution effects in the ^{222}Rn concentrations found.

3. Occupational radiological impact

The magnitude and behavior of the radionuclides involved in the PA production process revealed the need to determine its dosimetric impact on workers. Besides, a detailed dosimetric evaluation was carried out in the MAP and in the DAP plants located in the Huelva fertilizer facilities.

3.1. PA plants

Relatively small increments of the instantaneous external exposure were determined in the milling area during the processing of Moroccan phosphate rock. Only relatively higher values were found at or inside the silos used for the storage of the PR (0.25–0.37 mSv h^{-1}), due to the presence of large amounts of Moroccan phosphorite containing high activity concentrations of ^{238}U (around 1 kBq g^{-1}). It is worthy to note that lower values were also measured inside the rest room, room where the workers spend the majority of the time in the area.

Relatively high instantaneous external exposures over background were determined in some locations of the acid storage facilities and the concentration areas. However, the PA acids contain negligible amounts of ^{226}Ra and daughters, which are the main gamma emitter radionuclides inside the process.

The highest total effective external dose rates were measured inside the filtration facility (reaching 1.1–1.4 μSv h^{-1}), around the third washing stage, being 0.1 μSv h^{-1} the natural background of Huelva province. The measurements carried out above the filter itself, being a zone with very low occupation factors, show high external dose rate. The remaining measures were performed along the perimeter of the filter, and lower values were obtained due to radiation attenuation by the metallic lateral walls of the filtration system. The values obtained over the filter are clearly higher than expected in most cases. It is necessary to take into consideration that the values registered at 1 m over a semi-infinite extension of PG hardly surpass 0.3 μSv h^{-1} [13]. Consequently, an additional source of gamma radiation must be present in the filtering area, being a major contributor to the effective dose rate and, additionally, it should be present regardless of the PR rock treated in the process. Therefore, the existence of a permanent radioactive contamination in some parts of the filtering equipment (scales) can be derived. Quite high ^{226}Ra activities were determined in these scales, supporting the previous statement. We can then conclude from the external dose rate measurement campaigns that, with the exception of the milling area where only PR is involved, the values obtained for the external exposures cannot be correlated to the amount of radioactivity fluxing by them. Thus, the major fraction of the external exposure rates measured is produced by a permanent radioactive contamination fixed on the pipes, deposits and filtering systems in the plant.

The particulate matter concentration in outdoor surface air inside the factories (average = 370 ± 10 μg m^{-3}), and its natural alpha emitter radionuclides concentrations were measured. Based on these data, the additional committed effective doses due to aerosols inhalation received by workers were estimated, founding that the increases of the committed doses ranged 10–100 times lower than the limit value fixed for the public (1 mSv yr^{-1}).

3.2. MAP and DAP plants

External radiation and inhalation of particulate matter were considered the main routes by which increments in the exposure of the workers can be produced, whereas ^{222}Rn determinations were also performed in both plants in order to evaluate the possible additional risks associated with the presence of this radionuclide in enhanced amounts.

Regarding the eternal gamma radiation, the values were low and always in the range 0–0.08 $\mu Sv\ h^{-1}$ over background (which is 0.09 $\mu Sv\ h^{-1}$). These low values are to be expected when it is considered that the raw materials and final products are only enriched in radionuclides which are mainly alpha-emitters, and therefore they are depleted particularly in ^{226}Ra and daughters, as well as in radionuclides from the thorium series, and ^{40}K, which in NORM facilities represent the main contributors to the external gamma radiation.

With the external gamma dose rate determinations performed, it was possible to carry out a detailed assessment of the occupational exposures due to external radiation received by the workers carrying out their activities in the MAP and DAP plants.

The occupational exposures due to external radiation for different working categories were determined. The external exposures associated with the various kind of workers are relatively uniform (maximum of 0.1 $mSv\ yr^{-1}$), independent of the activities performed by each worker inside the plants, and are clearly below 1 $mSv\ yr^{-1}$.

The low external occupational exposures could be considered representative of the found ones for other plants devoted to the production of ammonium phosphate fertilizers that use phosphoric acid as a raw material. However, higher values can be expected in plants devoted to the production of NPK fertilizers that use MAP as a raw material, due to the contribution arising mainly from the ^{40}K intrinsically present in the natural potassium used for their generation.

The radiometric determinations performed in the aerosol filters collected in the plants show enhanced (and variable in time) mass activity concentrations of the radionuclides belonging to the uranium series in relation to the concentrations determined in the background area (El Arenosillo). However, no enhancements were observed for the radionuclides from the thorium series or for ^{40}K. In all cases these committed effective doses are moderate and remain below 0.12 $mSv\ yr^{-1}$.

The radiological study allowed us to conclude that the occupational doses received by the various worker categories in the plants do not reach the value of 0.3 $mSv\ yr^{-1}$ when considering the worst scenarios and all the possible pathways of exposure. The dose increments due to the external radiation and inhalation of particulate matter are the main routes of exposure. No single action or adoption of any radiation protection measure needs to be taken in order to protect the health of the workers.

Furthermore, the ^{222}Rn concentrations inside the two analyzed plants are also of no concern since they were around the 20 $Bq\ m^{-3}$, which is one order of magnitude lower than the established limit that would trigger the adoption of remediation actions.

4. Public and environmental radiological impact

Very high levels of different radionuclides from the uranium series were observed in waters (dissolution and suspended matter) and superficial sediments before the beginning of the close circuit of the pumping PG waters (1998), finding the highest ones in the Odiel river due to the direct releases of PG in this zone, since about 20% of the total PG production was directly released into Odiel Channel [27].

The sampling campaigns carried out before the change in the PG management policy (1998 year), the ^{226}Ra activity concentrations in the estuarine waters collected in the surroundings of the PA factories were around 10^2 Bq L^{-1} (two orders of magnitude higher than natural seawaters), diminishing clearly after 1998 to values near to the expected background ones.

The historical radioactive impact caused by the operation of the phosphoric acid plants in the neighboring estuaries have been found in selected sediment cores collected in the area. As an example, in the **Figure 3** is shown the ^{226}Ra profile obtained in one sediment core collected in the year 2000 in the Odiel estuary, together with the profile obtained simultaneously for ^{137}Cs. Assigning the maximum of the ^{137}Cs to the year 1963, it is possible to observe how the ^{226}Ra activity concentrations start to increase in layers with an age compatible with the starting of the operations of the factories in 1968, and how the levels of ^{226}Ra in different layers are clearly enhanced since 1968 until the surface layer of the sediment core [15].

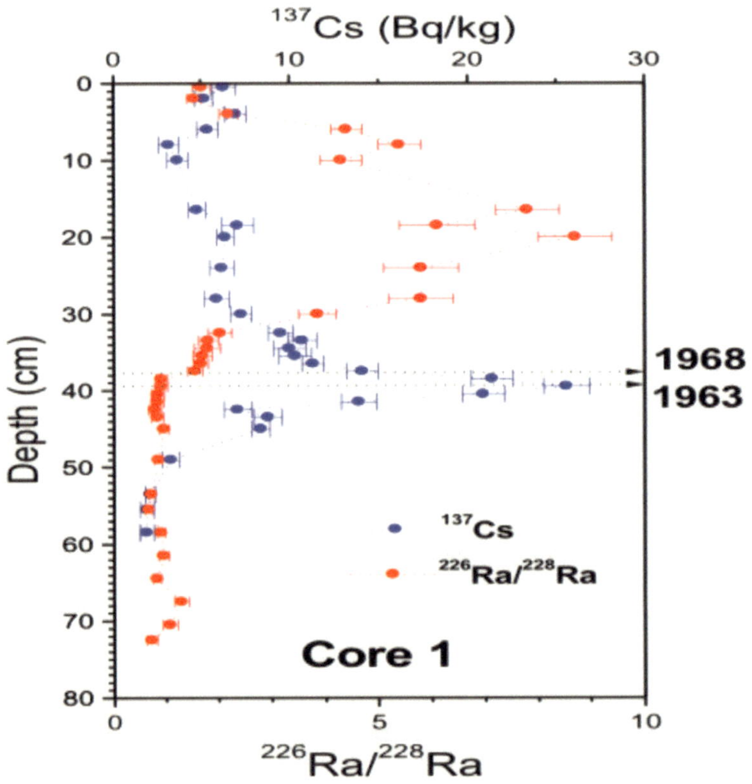

Figure 3. $^{226}Ra/^{228}Ra$ and ^{137}Cs activity profiles in one sediment core collected in the Odiel estuary. The results of ^{226}Ra are normalized to ^{228}Ra in order to discard the influence of possible changes in the composition of the sediments.

http://dx.doi.org/10.5772/intechopen.68567

The decreases observed for ^{226}Ra in the waters of the Odiel estuary after 1998 where found also in the Tinto estuary waters and for other radionuclides of the uranium series, indicating that after the adoption of the new management policy for the PG disposal in 1998, the leakage waters coming from the zone A of the PG piles, and the diffused escapes from zone B have a quite limited impact on the neighboring environment. These leakage and diffuse waters contain enhanced levels of radionuclides from the uranium series, but its volumetric magnitude is very low comparable with the water fluxes discharging from the rivers in the estuary.

Since 1998, only some sporadic increments were observed in the levels of ^{226}Ra in waters of the Odiel river, mostly due to the resuspension of the oldest sediments contaminated before 1998 [24]. In the waters and superficial sediments of the whole estuary, since 1998 a clear decrease in the levels of radionuclides such as ^{226}Ra and ^{210}Pb has been observed that nowadays are approaching natural background references [28].

After 1998, only one exception was found in relation with the general trend of lower activity concentrations in the different environmental compartments of the estuary. This exception corresponds to the Uranium. The temporal evolution of the U activity concentrations determined in several sampling points of the Tinto estuary was studied, founding that the levels of U in the waters after 1998 were similar or even higher than in the sampling campaigns carried out before.

The particular U results are mainly reflected in the waters and sediments of the Tinto estuary and can be explained by the existence of an additional source of this element reaching the estuaries. The Odiel and Tinto rivers have been historically, since the Roman Empire, mining rivers affected by the activities performed by the extraction of metals such as Cu in their margins. Both rivers are along their course affected by acid mining drainage, showing its waters pH values clearly acidic. The mining drainage waters provoke the preferential dissolution of the uranium from the mine substratum, these being drainage waters incorporated to river water flow and transported to the estuary where the associated U in dissolution experiment and its precipitation being associated to the change of pH provoked by the mixing of the acid waters transported by the river with the sea water. In relation specifically with U, its activity concentration in the Odiel and Tinto river estuaries, with independence of the restoration plan applied to the area, will not reach background values due to this additional historical source.

The importance of the mining activities in the upper margin of the rivers, and the influence of the acid mining drainage effect in the U levels in the studied estuaries is clearly reflected in the U profile obtained in one sediment core collected during 1998 in the Tinto estuary just in front of zone A of the PG piles. A modified version of the ^{210}Pb-dating method was applied for dating this sediment core. The U profile obtained reflects clearly the history of the mining activities in the area: start to increase sharply in the middle of the nineteenth century (when the mining activities started in great extent), approaching the maximum levels in the first century of the twentieth century when the mining activities in the area reach the maximum levels. Just after the Spanish civil war in the 1930s, the mining activities decrease as it is reflected in the decrease of the U levels found in the sediment layers formed during this time, suffering only a modest final increase since 1968 due to the influence of the phosphoric acid plants.

The peculiarities of the estuary formed by the confluence of the Tinto and Odiel rivers from the radioactive point of view, with relatively well-defined sources of several radionuclides from the uranium series, has converted the area in a natural laboratory where different radioecological studies have been and can be performed. In this sense, determinations of distribution coefficient factors (k_d), and of transfer factors soil to plants (TF), have been performed. In addition, this estuarine ecosystem and the huge amount of radioactive data generated in it, has served for the historical development of numerical hydrodynamic models to explain and predict the dispersion of the radioactive pollution and to gain in this way a lot of information about the environmental behavior of different radionuclides in estuarine compartments [29]. These models solve water circulation due to tides and river stream flows, include uptake/release reactions of radionuclides between the dissolved phase and bed sediments in a dynamic way (using kinetic transfer coefficients) and allow to evaluate the relative significance of the different radionuclides sources into the estuary as well as the independent effect of the two components of water circulation (tides are river flows) on radionuclide dispersion patterns.

Quite recently [19], a detailed study has been performed analyzing natural radionuclide concentrations in three species of halophytes plants growing in salt marshes from the Tinto estuary. The results of this work have shown that, (a) natural radionuclide concentrations (specially the U-isotopes) in the Tinto salt-marsh sediments are one order of magnitude higher than those determined in unpolluted salt marshes taken as a reference, and (b) these radionuclide enhancements are reflected in the different organs of the plants having increased concentration of these radionuclides. The transfer factor (TF) of the most polluted radionuclides (U-isotopes and ^{210}Po) in the Tinto salt marshes are one order of magnitude higher than in the unpolluted reference site indicating that the fraction of each radionuclide in the sediment originating from the pollution is more available for the plants than the indigenous fraction. This last conclusion allows indicating that the plants of the salt marshes are unhelpful as bioindicators or for the phytoremediation of radionuclides. In this work additional information was obtained by analyzing the internal distribution of the natural radionuclides in the plants, by distinguishing among its main organs (roots, stems, and leaves).

In representative aliquots of living biota that can be found in the Huelva estuary, natural radionuclide determinations have been also performed. In particular, shrimps and different shellfish species collected in the Huelva estuary have been analyzed after 1998, and it is found that the levels of ^{210}Po, ^{210}Pb, U-isotopes, and ^{230}Th are similar to the determined ones for the same species in aliquots collected in an unperturbed estuary located nearby, and to the determined ones for the same species in unperturbed estuarine ecosystems distributed over the world. This fact indicates that the radioactive pollution existing in the Huelva estuary, which is diminishing since 1998 to levels near to background nowadays, have a quite or very limited impact in the upper levels of the trophic chain.

If in addition, we have in consideration that (a) the possible interaction of the radioactive pollution existing in the estuary with continental waters is limited to a possible interaction with an aquifer located below the phosphogypsum piles that is saline and consequently its waters have no use in agriculture and as a source of drinking water, and (b) no detectable increments in particulate matter and ^{222}Rn activity concentrations have been observed in the vicinity of the PG piles, we can conclude indicating that the public radiological impact associated to the radioactive

http://dx.doi.org/10.5772/intechopen.68567

pollution of the Huelva estuary and to the presence of the PG piles is extremely limited—lower than 20 μSv yr^{-1}, being adopted for this estimation as extremely conservative assumptions.

Acknowledgements

Parts of this chapter are reproduced from the authors' previous publications (Refs. [10, 11, 14], among others). This work was partially supported by the Ministry of Economy and Competitiveness of Spain through the research project with Ref.: CTM2015-68628-R.

Author details

José Luis Guerrero-Márquez[1], Fernando Mosqueda Peña[1], Juan Mantero[2], Guillermo Manjón[2], Rafael García-Tenorio[2] and Juan Pedro Bolívar[1]*

*Address all correspondence to: bolivar@uhu.es

1 Department of Integrated Sciences, Campus El Carmen, University of Huelva, Huelva, Spain

2 Department of Applied Physics II, University of Seville, Seville, Spain

References

[1] Northolt AJG, Sheldon RP, Davidson DF, editors. Phosphate Deposits of the World. Vol. 2: Phosphate Rock Resources. Cambridge, UK: Cambridge University Press; 2005. p. 600

[2] International Fertilizer Industry Association. Phosphate production in the new economics. Presentation to Joint IAEA-FIPR Phosphate Workshop. Bartow, Florida. 2006

[3] Fertiberia S.A. Documentation on the Industrial Process at the Huelva Phosphoric Acid Production Plant. Spain: Fertiberia, SA; 2002 (in Spanish)

[4] Abril JM, García-Tenorio R, Enamorado S, Hurtado MD, Andreu L, Delgado A. The cumulative effect of three decades of phosphogypsum amendments in reclaimed marsh soils from SW Spain: ^{226}Ra, ^{238}U and Cd contents in soils and tomato fruit. Science of the Total Environment. 2008;**403**(1-3):80-88. DOI: 10.1016/j.scitotenv.2008.05.013

[5] Borrego E, Mas JL, Martín JE, Bolívar JP, Vaca F, Aguado JL. Radioactivity levels in aerosol particles surrounding a large TENORM waste repository after application of preliminary restoration work. Science of the Total Environment. 2007;**377**(1):27-35. DOI: 10.1016/j.scitotenv.2007.01.098

[6] Periáñez R, Absi A, Villa M, Moreno HP, Manjón G. Self-cleaning in an estuarine area formerly affected by ^{226}Ra anthropogenic enhancements: Numerical simulations. Science of the Total Environment. 2005;**339**(1-3):207-218. DOI: 10.1016/j.scitotenv.2004.07.029

[7] OSPAR Commission. Discharges of radioactive substances into the maritime area by non-nuclear industry. Radioactive Substances Series. 2002. ISBN 0-946956-90-1. P. 60

[8] OSPAR Commission. Liquid discharges from nuclear installations in 2007. Radioactive Substances Series. 2009. ISBN 978-1-906840-96-9 Publication Number: 456/2009. P. 24

[9] Absi A, Villa M, Moreno HP, Manjón G, Periañez R. Self-cleaning in an estuarine area formerly affected by ^{226}Ra anthropogenic enhancements. Science of the Total Environment. 2004;**329**(1-3):183-195. DOI: 10.1016/j.scitotenv.2004.03.001

[10] Bolívar JP, Martín JE, García-Tenorio R, Pérez-Moreno JP, Mas JL. Behaviour and fluxes of natural radionuclides in the production process of a phosphoric acid plant. Applied Radiation and Isotopes. 2009;**67**(2):345-356. DOI: 10.1016/j.apradiso.2008.10.012

[11] Bolívar JP, Pérez-Moreno JP, Mas JL, Martín JE, San Miguel EG, García-Tenorio R. External radiation assessment in a wet phosphoric acid production plant. Applied Radiation and Isotopes. 2009;**67**(10):1930-1938. DOI: 10.1016/j.apradiso.2009.06.004

[12] Lozano RL, Bolívar JP, San Miguel EG, García-Tenorio R, Gázquez MJ. An accurate method to measure alpha-emitting natural radionuclides in atmospheric filters: Application in two NORM industries. Nuclear Instruments and Methods in Physics Research Section A: Accelerators, Spectrometers, Detectors and Associated Equipment. 2011;**659**(1):557-568. DOI 10.1016/j.nima.2011.08.006

[13] Mas JL, Bolívar JP, García-Tenorio R, Aguado JL, San Miguel EG, González-Labajo J. A dosimetric model for determining the effectiveness of soil covers for phosphogypsum waste piles. Health Physics. 2001;**80**(1):34-40. DOI: 10.1097/00004032-200101000-00007

[14] Gázquez MJ, Mantero J, Mosqueda F, Bolívar JP, García-Tenorio R. Radioactive characterization of leachates and efflorescences in the neighbouring areas of a phosphogypsum disposal site as a preliminary step before its restoration. Journal of Environmental Radioactivity. 2014;**137**:79-87. DOI: 10.1016/j.jenvrad.2014.06.025

[15] San Miguel E, Pérez-Moreno JP, Bolívar JP, García-Tenorio R. A semi-empirical approach for determination of low-energy gamma-emmiters in sediment samples with coaxial Ge-detectors. Applied Radiation and Isotopes. 2004;**61**(2-3):361-366. DOI: 10.1016/j.apradiso.2004.03.009

[16] Aguado JL, Bolívar JP, García-Tenorio R. Sequential extraction of ^{226}Ra in sediments from an estuary affected historically by anthropogenic inputs of natural radionuclides. Journal of Environmental Radioactivity. 2004;**74**(1-3):117-126. DOI: 10.1016/j.jenvrad.2004.01.016

[17] Martínez-Aguirre A, García-León M. ^{210}Pb distribution in riverwaters and sediments near phosphate fertilizer factories. Applied Radiation and Isotopes. 1996;**47**(5-6):599-602. DOI: 10.1016/0969-8043(96)00002-4

[18] Martínez-Aguirre A, Periañez R. Soil to plant transfer of 226Ra in a marsh area; modelling application. Journal of Environmental Radioactivity. 1998;**39**:199-213. DOI: 10.1016/S0265-931X(97)00048-9

[19] Luque CJ, Vaca F, García-Trapote A, Hierro A, Bolívar JP, Castellanos EM. Radionuclides transfer into halophytes growing in tidal salt marshes from the Southwest of Spain. Journal of Environmental Radioactivity. 2015;**150**:179-188. DOI: 10.1016/j.jenvrad.2015.08.002

[20] Pérez-Moreno JP, San Miguel EG, Bolívar JP, Aguado JL. A comprehensive calibration method of Ge detectors for low-level gamma-spectrometry measurements. Nuclear Instruments and Methods in Physics Research Section A: Accelerators, Spectrometers, Detectors and Associated Equipment. 2002;**491**(1-2):152-162. DOI: 10.1016/S0168-9002(02)01165-8

[21] Lehritani M, Mantero J, Casacuberta N, Masqué P, Garcia-Tenorio R. Comparison of two sequential separation methods for U and Th determination in environment samples by alpha particle spectrometry. Radiochimica Acta. 2012;**100**(7):431-438. DOI: 10.1524/ract.2012.1933

[22] Hallstadius L. A method for the electrodeposition of actinides. Nuclear Instruments and Methods in Physics Research. 1984;**223**:457-462. DOI: i:10.1016/0167-5087(84)90659-8

[23] UNSCEAR. Report of the United Nations Scientific Committee on the Effects of Atomic Radiation. New York: United Nations; 2000

[24] Hierro A, Bolivar JP, Vaca F, Borrego J. Behavior of natural radionuclides in surficial sediments from an estuary impacted by acid mine discharge and industrial effluents in Southwest Spain. Journal of Environmental Radioactivity. 2012;**110**:13-23. DOI: 10.1016/j.jenvrad.2012.01.005

[25] Bolívar JP, García-Tenorio R, García-León M. Radioactive impact of some phosphogypsum piles in soils and salt marshes evaluated by γ-ray spectrometry. Applied Radiation and Isotopes. 1996;**47**(9-10):1069-1075. DOI: 10.1016/S0969-8043(96)00108-X

[26] Abril JM, García-Tenorio R, Manjón G. Extensive radioactive characterization of a phosphogypsum stack in SW Spain: ^{226}Ra, ^{238}U, ^{210}Po concentrations and ^{222}Rn exhalation rate. Journal of Hazardous Materials. 2009;**164**(2-3):790-797. DOI: 10.1016/j.jhazmat.2008.08.078

[27] Bolívar JP, García-Tenorio R, García-León M. Enhancement of natural radioactivity in soils and salt-marshes surrounding a non-nuclear industrial complex. Science of the Total Environment. 1995;**173-174**:125-126. DOI:10.1016/0048-9697(95)04735-2

[28] Villa M, Mosqueda F, Hurtado S, Mantero J, Manjón G, Periañez R, Vaca F, García-Tenorio R. Contamination and restoration of an estuary affected by phosphogypsum releases. Science of the Total Environment. 2009;**408**(1):69-77. DOI: 10.1016/j.scitotenv.2009.09.028

[29] Periáñez R, Hierro A, Bolívar JP, Vaca F. The geochemical behavior of natural radionuclides in coastal waters: A modeling study for the Huelva estuary. Journal of Marine Systems. 2013;**127**:82-93. DOI: 10.1016/j.jmarsys.2012.08.001

Chapter 6

Phosphate-Based Organic-Inorganic Hybrid Materials: Design, Construction and Technological Applications

Hussein M. H. Alhendawi, Ernesto Brunet and Elena Rodríguez-Payán

Additional information is available at the end of the chapter

http://dx.doi.org/10.5772/68140

Abstract

The art of designing and synthesizing organic molecules has reached very high levels of sophistication, based on a *relatively simple set of rules* that guide both the invention and synthesis of new compounds. This set of rules is construed as the *rational synthetic method* of organic chemistry. As material chemists confronted to the task of building new solid structures with tailored chemical properties, we do inevitably need to develop some rational approach and to establish the corresponding set of rules allowing a realistic level of predictive knowledge in the construction of solid scaffolds. These conditions are reasonably accomplished by the use of layered salts of tetravalent transition metals, namely zirconium phosphate (ZrP). The placing of organic molecules between the layers of ZrP is quite straightforward, can easily be controlled and leads to enduring, solid materials where the confinement makes the organic molecules to show new properties at the supramolecular level. The chemistry of metal phosphates/phosphonates will be detailed in relation with the following topics: (i) molecular recognition, (ii) chemically driven porosity changes, (iii) chiral memory and supramolecular chirality, (iv) luminescence signalling, (v) photo-induced electron-transfer processes, (vi) hydrogen storage, (vii) confinement of drugs and (viii) metal uptake.

Keywords: hybrid-layered materials, zirconium phosphate, porous materials, intercalation reactions, topotactic exchange, hydrogen storage, luminescence signalling, supramolecular chirality, molecular confinement, metal uptake

1. Introduction

Phosphoric acid (PA) is an important industrial acid that is widely used to manufacture a myriad of different industrial products for an extensive range of applications. The exploitation of

PA has become global and its utilization to producing innovative chemical structures is a trend that is expanding worldwide as well. Along these lines, in this chapter, we focus on our past and ongoing research devoted to the rational design and construction of *porous frameworks* based on layered metal phosphates and their technological applications.

Phosphate-based-layered organic-inorganic materials, especially those built from octahedral corner-sharing zirconium and tetrahedral phosphate, have attracted much attention and aroused increasing interest from the viewpoint of developing novel functional porous materials for many fields of *modern technology*, such as luminescence materials [1], chiral molecular selectors [2, 3], hydrogen adsorbents [4–6], drug-delivery systems [7] and materials for waste water treatment [8–10].

2. Layered metal (IV) phosphate

The area of chemistry that explores the porous materials has received a great impulse in the past 40 years. In addition to the classic porous materials (e.g. activated charcoal, silica, alumina and zeolites), synthetic materials have been developed with analogous structures to those observed in Nature. An important landmark in this field has been the preparation of synthetic zeolites, which have a significant role as they could be used as catalysts, adsorbents and ion exchangers.

An important effort has been dedicated to the development of other porous materials with analogous three-dimensional (3D) structures. To this effect, the chemistry of metallic phosphates [11] with laminar structure has received a great impulse, because they are very versatile materials whose handling complies with what we have termed rational synthetic method, that is, the development of a set of relatively simple rules that confers sufficient predictive knowledge to the building of crystalline materials. The synthetic rationale elaborated by us through a number of years [12] is modular because it comprises the design and synthesis of appropriate organic molecules in one hand, and on the other their stepwise introduction into the inorganic framework. Common characteristics of this synthetic approach to solid materials based on layered metallic phosphates are as follows: (i) their reactions proceed at low temperatures; (ii) they can be functionalized with either organic or inorganic molecules, by ionic or covalent bonds, which confers them with very specific properties and a broad field of application; (iii) their functionalization is accomplished by means of *topotactic* reactions, which do not affect the fundamental structure of the material, so that it is possible to predict the arrangement of the final products; (iv) when the introduced organic moieties have two reacting ends, it is possible for them to bind consecutive lamellae of these materials and three-dimensional, pillared structures can be thus easily created.

Basically, layered structures formed by zirconium phosphate (ZrP) are produced by octahedra-tetrahedra combinations in two main different formats, named α- and γ-phases, where zirconium coordination varies from one to another (**Figure 1**). In both α- and γ-structures, zirconium coordinates octahedrically to six oxygens of six surrounding phosphates. However, in α-phase, all phosphates (HPO_4) are equivalent and they all use three oxygens to bond to zirconium, whereas in the γ-phase phosphates are of two types: PO_4 that uses its four oxygens to

http://dx.doi.org/10.5772/68140

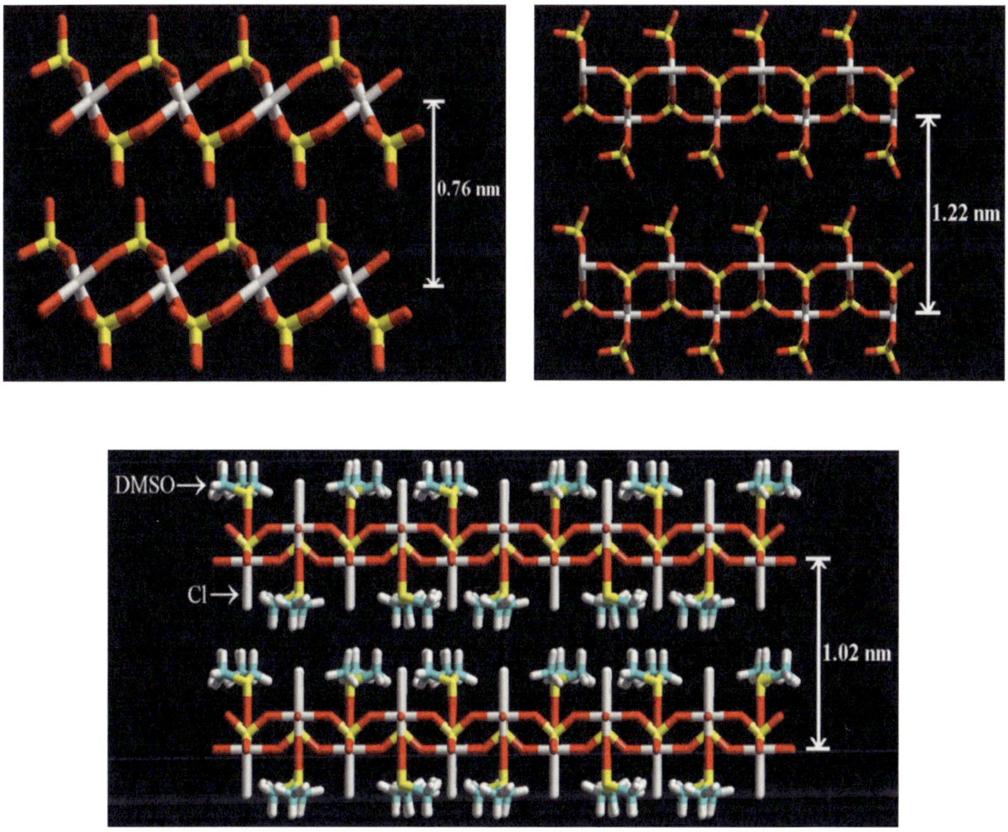

Figure 1. Tube models (Zr, grey; P, yellow; O, red; H, white) of two consecutive layers of α-ZrP (top left), γ-ZrP (top right) and λ-ZrP (bottom). The experimental interlayer distances are shown (Color figure online).

coordinate to zirconium and H_2PO_4 that puts into play only two oxygens to do that. Thus, the molecular formulae for α- and γ-ZrP are $Zr(HPO_4)_2{\cdot}H_2O$ and $Zr(PO_4)(H_2PO_4){\cdot}2H_2O$, respectively [13–15].

A third type of ZrP (hereafter referred as λ-ZrP) structurally related to γ-ZrP can be obtained by coordinating the four oxygen atoms of the PO_4 group to four different zirconium atoms and completing the zirconium octahedral structure with one negatively charged monodentate ligand (usually chloride or fluoride ion) and another neutral monodentate ligand (dimethylsulphoxide) [13–15]. The structure of λ-ZrP [$ZrPO_4Cl(dmso)$] is shown in **Figure 1**.

3. Reactions

3.1. Intercalation

The intercalation properties of zirconium phosphate have been extensively investigated due to their interesting fundamental aspects and possible applications as well, including ion exchange, high-temperature stable organic-inorganic composites, ion conduction, catalysis, and so on [16–18].

It is well known that the intercalation chemistry of α- and γ-ZrP is dominated by the presence in the interlayer region of Brønsted acid OH groups [19]. They are powerful phases for the entrapping and exchanging of organic cations such as primary alkylamines and aminophenyl- and pyridinium-substituted porphyrins among many other examples [8, 9, 19].

We have recently prepared λ-ZrP with Brønsted acid properties, λ-$ZrPO_4[OOC\text{-}(CH_2)_4\text{-}COOH](CH_3)_2SO$ (**Figure 2**), by means of topotactic anion exchange of the chloride ligand in λ-zirconium phosphate [λ-$ZrPO_4Cl(CH_3)_2SO$] with adipic acid (AA). The so-synthesized λ-(AA)ZrP solid phase has a higher acidic character and a larger gallery height in comparison to the pristine λ-ZrP (1.47 and 1.02 nm, respectively). Therefore, this material is expected to be a suitable host for intercalation of relatively large basic guests [20].

With respect to the intercalation of primary alkylamines into layered ZrP, it has many potential uses and technological applications. Moreover, it constitutes a very powerful synthetic tool to facilitate the insertion of large organic and inorganic radicals inside the interlayer gallery of layered solid materials [21–31]. In this regard, it should be mentioned that the intercalation chemistry of λ-ZrP has not as yet been as well investigated as that of α- and γ-ZrP [21–29].

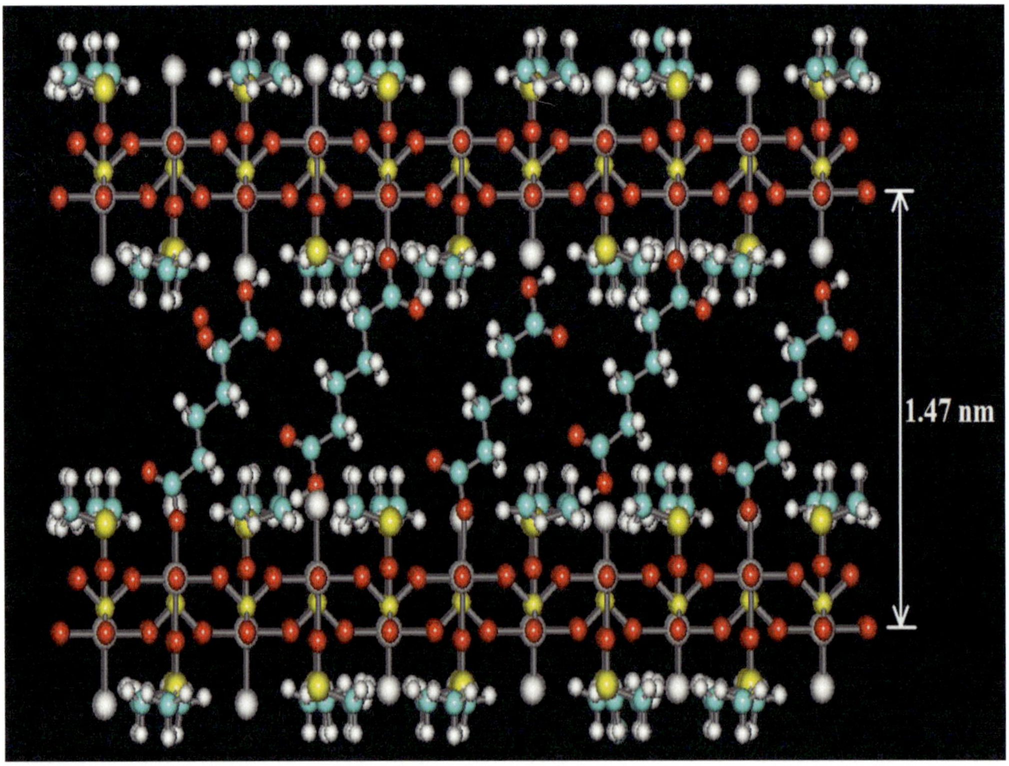

Figure 2. Possible arrangement of AA molecules in λ-(AA)ZrP complying with the observed interlayer distance.

Recently, primary alkylamines, namely methyl-, ethyl-, butyl- and hexylamine, has been successfully intercalated into λ-ZrP in our laboratory. In this way, λ-type materials with increasing interlayer separation are easily obtained [32].

The experimental data are only compatible with the formation of a double layer of alkylamines in a zig-zag conformation as shown in **Figure 3**. This behaviour is common to all ZrP phases, the relationship between interlayer distance and amine chain length being linear indicating that the longitudinal axis of the alkyl group should be inclined relative to the average plane of the inorganic layers [32].

The increase of basal spacing suffered by ZrP upon alkylamine intercalation makes it possible for the water molecules to penetrate the interlayer region to solvate the polar groups. The suspension of the resulting amine-intercalated materials in water may be further manipulated to carry out topotactic exchange reactions and insert large organic and inorganic species of different functionality (polar-non-polar, rigid-flexible, chiral-achiral, acid-base, etc.) inside the interlayer galleries of ZrP. Accordingly, new tailored organic-inorganic materials for many scientific applications can be obtained [33, 34].

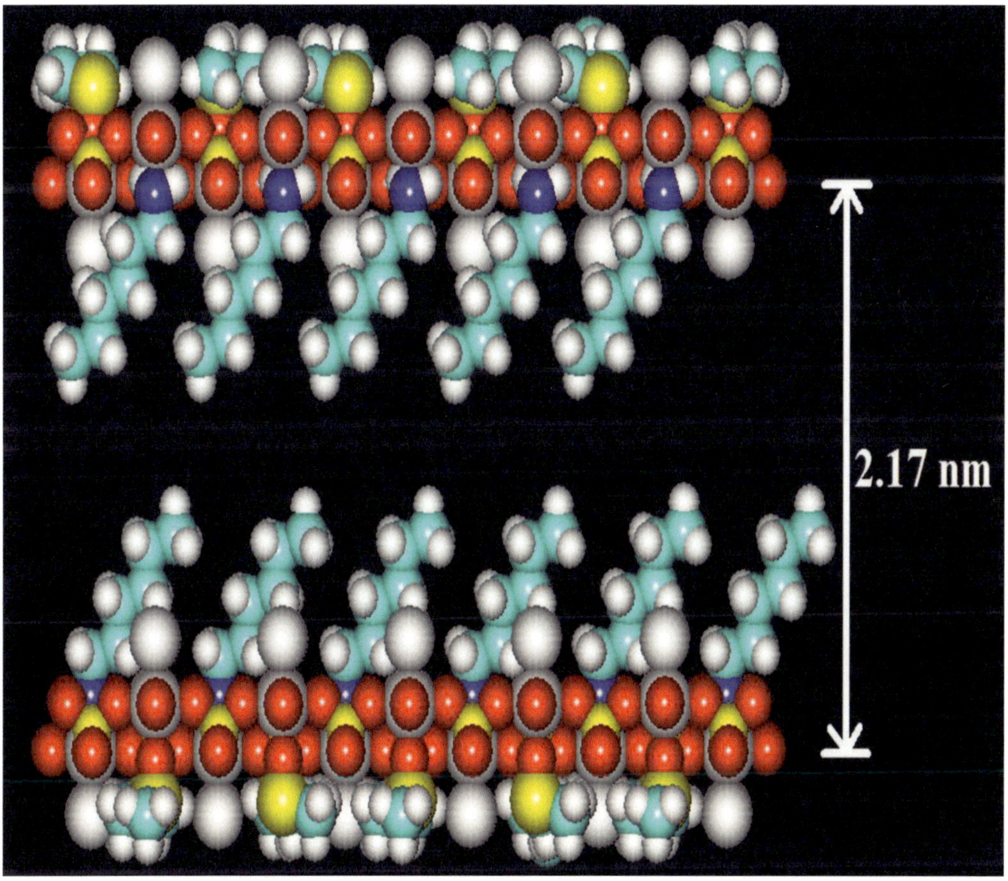

Figure 3. Possible arrangement of hexylamine inside λ-ZrP complying with the observed interlayer distances.

3.2. Topotactic exchange

The most interesting reaction of γ-ZrP is that in which the acidic surface phosphates are replaced by other phosphorous functions such as phosphonates (**Figure 4**) [26, 27]. For this reaction to take place at a reasonable rate, the layers of γ-ZrP have to be separated first in a process named exfoliation [22], either by intercalation of an alkylamine of sufficient chain length or by solvation of the inorganic layers in a 1:1 mixture of water-acetone at 80°C. The exchange reaction takes place in a topotactic manner, that is, it occurs without affecting the integrity and rigidity of the inorganic layers.

Mixed organic derivatives of γ-ZrP such as γ-ZrP/phosphonate, γ-ZrP/phosphite and γ-ZrP/phosphinate compounds are possible. Thus, γ-ZrP derivatives can be represented by the formula $ZrPO_4.O_2PRR'.nS$, where R and R′ can be H, OH or an organic group and S is the intercalated solvent.

When the topotactic phosphate/phosphonate exchange reactions occur at the same time on the facing surfaces of two adjacent layers by a molecule with two phosphonate groups, pillared γ-ZrP is formed (**Figure 4**).

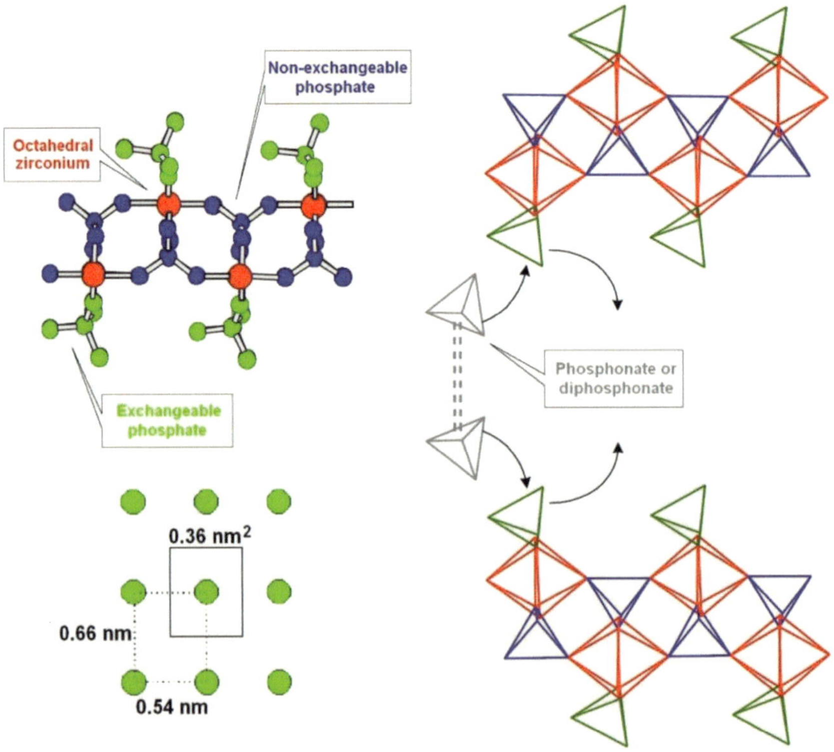

Figure 4. From top-left counterclockwise: (i) schematic structure of a portion of a layer of γ-ZrP; (ii) dimensions of the grid formed by superficial phosphates on a layer; (iii) representation of the topotactic exchange of γ-ZrP by a phosphonate or diphosphonate.

Regarding the pillared derivatives of α-ZrP, they are obtained by allowing zirconium fluoro-complexes to decompose in the presence of diphosphonic acids [26, 27].

If the pillars have a suitable height and are sufficiently spaced, microporous materials are thus obtained. Pillars could be either rigid or non-rigid, the latter being the most versatile as the porosity of the materials can be controlled by simple chemical means [35–38]. The outstanding features of these materials come from their ability to combine the properties of both the organic pendant groups with those of the inorganic host [39]. Such an approach can result in the design and development of hybrid organic-inorganic materials with tailor-made properties. These materials can function as molecular sieves of controlled pore size [39], shape-selective catalysts [40, 41], molecular sorbents [42, 43] and stationary phases for chiral molecular recognition [2, 44].

Concerning the relatively unexplored λ-ZrP phase, it is worth mentioning that the chloride and dimethyl sulphoxide ligands of λ-ZrP (*cf.* **Figure 1**) can be topotactically exchanged by other monovalent anionic ligands (alkoxides, carboxylates, etc.) and neutral monodentate ligands (sulphoxides, amines, etc.), respectively [45]. If dialkoxides or dicarboxylate ligands are used, λ-layered-pillared organic-inorganic derivatives are obtained.

We have recently prepared several new porously dynamic layered-pillared materials (**Figure 5**) with a general formula of λ-$ZrPO_4(OH)_{1-x}(OOC(CH_2)_nCOO)_{x/2}$(dmso) ($n$ = 6, 8 and 10) by post-synthesis modification of the inorganic layers of λ-zirconium phosphate (λ-ZrP), where the superficial chloride monovalent anionic ligands of λ-layer are partially exchanged with the divalent anionic ligands of a series of long-chain aliphatic dicarboxylic acids, namely octane-dioic, decanedioic and dodecanedioic acids.

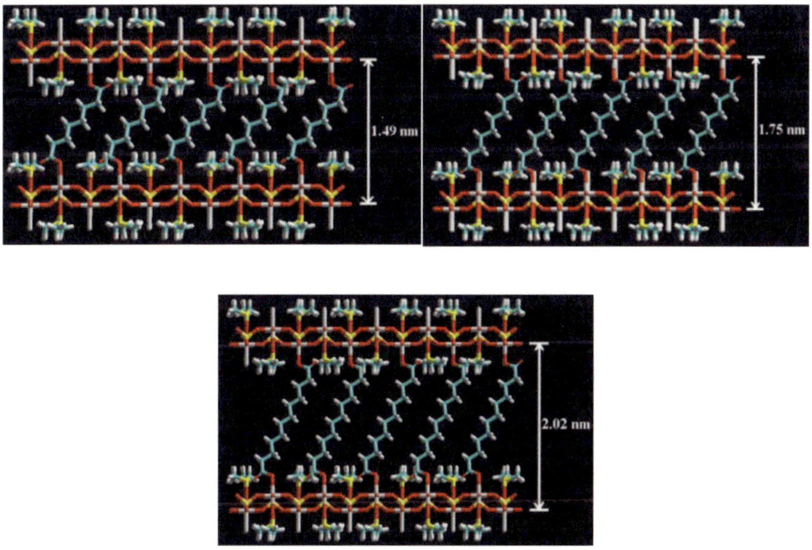

Figure 5. Possible arrangement of octanedioic acid (*top left*), decanedioic acid (*top right*) and dodecanedioic acid (*bottom*) inside λ-ZrP.

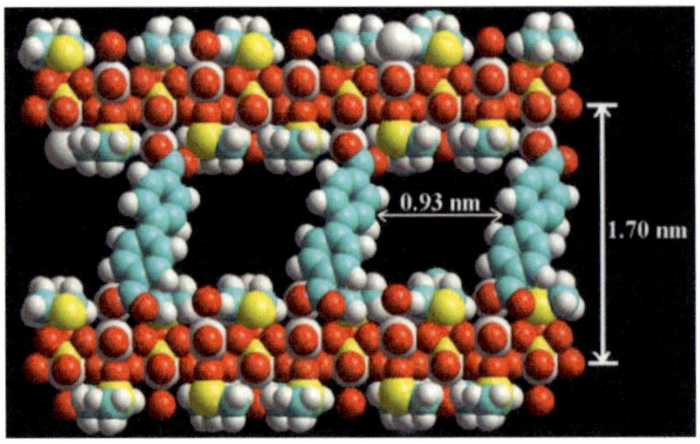

Figure 6. Possible arrangement of bpdc inside λ-ZrP (40% pillaring).

More recently, a new λ-type-layered-pillared rigid microporous framework based on λ-ZrP and 1,4-biphenyldicarboxylate (bpdc) (**Figure 6**) has been prepared in our laboratory [46].

4. Design, processing and applications of ZrP-based porous materials

Over the past decades, the chemistry of ZrP structures has been studied intensively [12]. In particular, compounds with organic radicals covalently attached to the inorganic layers of ZrP. These organic-inorganic frameworks are prepared by two main synthetic approaches: (i) thermal decomposition of zirconium fluoro-complex in the presence of phosphonic acids (α-type derivatives) [19, 24, 26, 27, 35], and (ii) post-synthetic modification of the of inorganic surface of γ- or λ-ZrP by means of topotactic ligand exchange reactions with phosphonic acids [5, 19, 24, 26, 27, 47] or carboxylic acids [32, 33, 45], respectively.

It is important to stress out that the post-modification of the inorganic surface of γ- or λ-ZrP is carried out without the destruction of the former framework structure (**Figure 4**). Therefore, protocols and strategies of rational synthetic design can be easily applied which turn ZrP into a *chemical-engineering tool* to build organic-inorganic frameworks in the solid state. This rationale provides the opportunity to effectively adjust the relative position of the organic radicals inside the solid matrix and thus adjust the latter's molecular geometry as a whole. As a result, the *by and large properties* of the final material are tuned with high precision [12]. At this point, the reader should not be misled by the conceptual, apparent simplicity of the topotactic exchange reactions. In particular, the mechanism of the replacement of the surface phosphates in γ-ZrP must imply a phenomenal process, even though it looks schematically simple. Unfortunately, there are no studies in the literature concerning the thermodynamic parameters of this process, probably because the colloidal milieu, in which the reaction is performed, makes the measurements difficult.

Considering versatility, the lambda structure is special. As we mentioned earlier, a layer of phosphates sustains the zirconium metals whose apical bonds are bonded to negative and neutral basic species, both pointing to the interlayer space (bottom of **Figure 1**). In the starting lambda structure, each zirconium is typically bonded to chloride anion and neutral

http://dx.doi.org/10.5772/68140

dimethylsulphoxide molecule [48, 49] which, in principle, can be replaced by any organic chemical structure, that is, carboxylates and amines, in a topotactic manner [32–34, 45]. Recently, we have prepared several new soft and rigid phosphate-based λ-type organic-inorganic two-dimensional (2D) and 3D frameworks with *variable porosity* [10, 20, 32, 34].

In summary, the chemical properties of the ensuing composites can be tailored by carefully selecting the physical and chemical functionalities of the organic building blocks that will be modified by their confinement between the inorganic layers. We envisage that the only limitation of this entire approach is the imagination and needs of the researcher.

4.1. Variable porosity

The realization that the layers of ZrP phases containing organic derivatives can be considered as a multiple-storey labyrinth immediately hits the idea of molecular recognition. In principle, if one was able to control the level of incorporation of organic molecules within the layers, the resulting material would be porous enough to allow guest molecules to travel between the lamellae and differentially interact with them. One way of achieving this goal is to include crown ethers.

Our seminal paper on the topotactic exchange with various crown ethers [50] of different sizes, containing one or two phosphonate groups, allowed for a higher control of the number of incorporated species, giving rise to highly porous materials. Yet, despite all the interesting structural findings, the true importance of the work is the achievement for the first time ever of multiple, sequential topotactic exchange reactions, namely the further replacement with methylphosphonate of the remaining phosphates from the initial, partial phosphate/crown-ether phosphonate exchange reaction.

Another way of approaching the building of crown-ether-like supramolecular structures is the use of diphosphonates derived from linear polyethylene glycol chains, in order to create pillared materials with polyethylenoxa chains (**Figure 7**). By simply varying the chain length, the access to a set of different labyrinths with sticky columns of different height and different porosity is made accessible. Further replacement of the remaining phosphates by hypophosphite (H_2PO_2) led to different sets of materials, named as *polar/polar* (polar columns and polar surface phosphates) and *polar/non-polar* (polar columns and non-polar surface hypophosphites) of the following general formula:

$Zr(PO_4)(H_2PO_4)_{0.75\text{-}x}[HO_3P(CH_2)_2(OCH_2CH_2O)_n(CH_2)_2PO_3H]_{0.125}(H_2PO_2)_x$ (n = 2–6; x = 0–0.75).

The unexpected behaviour occurred when these materials were treated in aqueous dispersion with the small base methylamine. Surprisingly enough, the *polar/polar* material steadily augmented its interlayer distance (*ca.* 1 nm) with the increasing amount of the small intercalating amine (pH = 2–9). Yet, its *polar/non-polar* counterpart suffered an abrupt increase of interlayer distance (*ca.* 1.4 nm) in a much narrower pH range (pH = 4–5). The starting interlayer distance at low pH was way shorter (1.3–1.7 nm) than that expected considering the length of the polyethylenoxa chains. Therefore, in the absence of methylamine, the chains had to arrange in a parallel fashion to the inorganic layers leading to a compact material. Nonetheless, at high pH the distance heavily increased reaching a value only compatible with extended, perpendicular polyethylenoxa columns. Yet, methylamine is so tiny that the substantial distance increment could not be solely attributed to the size of the intercalated amine. The plausible explanation could be that, owing to the intercalation with the basic methylamine molecules, hydrogen bonds between pillars (polyethylenoxide chains)

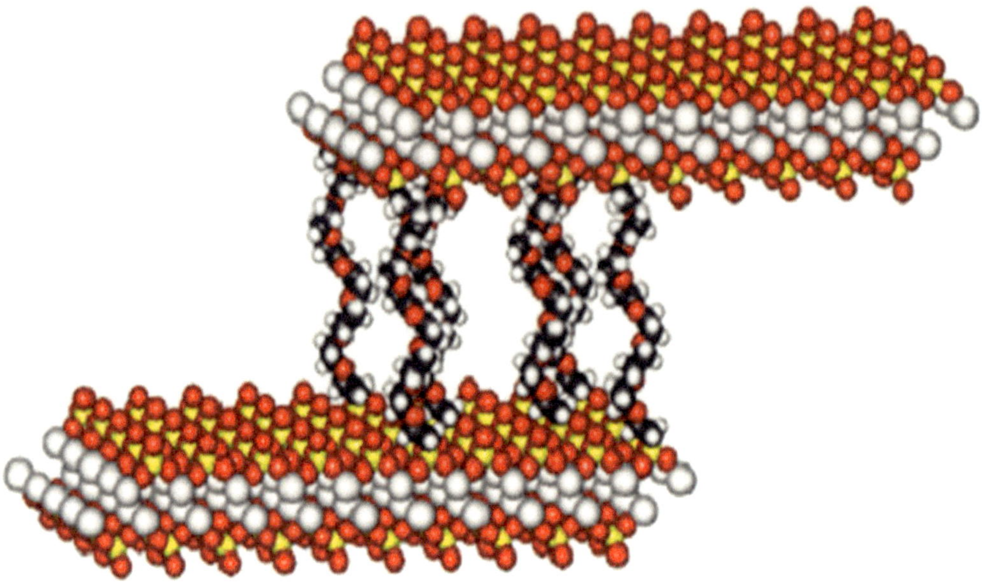

Figure 7. Idealized molecular model of γ-**ZrP** exchanged with pentaethylene glycol diphosphonate as a representative example.

and $O_2P(OH)_2$ acidic groups of the layers are broken. Therefore, the number of water molecules in the interlayer region is increased and the polyethylenoxide chains are forced to align perpendicular to the layers. This odd supramolecular behaviour in the solid state, where the porosity of a layered material is abruptly increased in response to a simple acid-base reaction in the solid-liquid interface, has never been observed before [44].

4.2. Dissymmetry and luminescence signalling

Molecular modelling of the *polar/polar* and *polar/non-polar* materials (see previous section) in the extended arrangement, that is, when the amine was present, suggested that in order to fit the largest achieved interlayer distances, the polyethylenoxa columns had to be helicoidally arranged. Polyethylene glycols are actually known to attain that conformation in aqueous solution. Yet, what makes a notable difference in the previously discussed materials is that the polyethylenoxa columns are covalently attached by both ends to the inorganic layers, thus conferring to the whole scaffold a supramolecular dimension. In principle, should the appropriate conditions be established, the random P/M helicity of the chains could be driven towards homochirality. Notably, this homochirality has been revealed by optical rotation measurements. To achieve that, we performed the following simple experiment.

Dispersions of small amounts in water/acetone of either native γ-**ZrP** or exchanged at 25% with hexaethylene glycol diphosphonate (n = 6; *cf.* molecular formula in previous section) displayed no sizeable optical rotation as expected. However, the material pillared with hexaethylene glycol diphosphonate intercalated with (+)-phenethylamine rendered a relatively large value of optical rotation. Surprisingly, when the enantiomerically pure amine was smoothly replaced with aquiral hexylamine, the optical rotation kept amazingly showing through. Nuclear magnetic resonance (NMR) experiments proved that the (+)-phenethylamine was no longer in the material. Hence, the only possibility left to explain the remaining optical activity must be laid upon the polyethylenoxa columns that had to acquire a certain bias in their initially random P/M helicity. This bias had to be induced by the concourse of the chiral (+)-phenethylamine and maintained when the latter was replaced by the achiral hexylamine. These facts were interpreted as an expression of supramolecular chiral memory [3]; no similar cases can be found in the literature.

The unexpected findings went a step further. What would happen if the chiral influence was intrinsic to the polyethylenoxa chains? To answer this question, enantiomerically pure diphosphonates were prepared [2] and topotactically exchanged into **γ-ZrP** at the 25% level. The intrinsically chiral polyethylenoxa columns were forced to acquire different conformations by the intercalation of amines of increasing length. The longer the amine, the more elongated the organic pillar. As anticipated, the basal spacing progressively increased with the length of the amine. On the contrary, it was outstandingly found that the optical rotation presented a maximum value when hexylamines were intercalated. Molecular modelling of the different intercalates, performed by maintaining the interlayer distance fixed at the corresponding experimental value and allowing the organic chain to freely reach the most stable conformation, showed that the maximum expression of helicity was precisely achieved at the spacing attained with butyl- or hexylamine. No other reasonable explanation could be found to account for the observed variation of the optical rotation versus interlayer distance.

In summary, the unexpected chirality behaviour of the accounted materials clearly pointed to the important fact that chiral properties can be created and amplified at the supramolecular level in the solid state and finely tuned by mild reactions in the solid-liquid interface [51–53].

Important research is being directed to the design of molecular systems able to display the strong luminescence of lanthanide metals [53]. To accomplish it with efficiency, two stringent conditions have to be met: (i) a suitable organic chromophore should absorb light and efficiently transfer energy to the metal (*antenna effect*), (ii) the coordination sphere of the metal should be free of water molecules because the OH oscillators easily quench metal emission. The first condition has been achieved either by simple acid-base intercalation or by topotactic exchange of suitable chromophores in the organic-inorganic lattice. The second, by using the pillared materials with polyethylenoxa columns described above, constituted an excellent spider web to enshroud the oxygenphylic lanthanide metals and isolate them within the solid matrix. Additionally, the chromophores have been provided with chirality. The resulting materials exhibited circularly polarized luminescence (CPL) of lanthanides, an elusive property in the solid state which could find important applications in the technology of optical handling of information.

4.3. Gas and vapour storage

Hydrogen storage is attained by *physisorption* in porous matrices [4, 54] among other procedures. The attachment of polyphenyl or polyphenylethynyl diphosphonates to either of the ZrP phases (cf. **Figure 6** as an example) led to materials with slit-like ultramicropores of different length. Further exchange reactions allowed for the attachment of polar or non-polar groups at both ends of the pores. This set of materials was tested for hydrogen storage. The best results were obtained with the material named **αT60Li**, α-ZrP with 60% of terphenyl-diphosphonates [6], the remainder phosphates bearing Li^+ as counterions. At 800 Torr and 77K, 1.7% w/w of hydrogen could be stored and the Department of Energy goal for 2010 (45 g of hydrogen per L of material) was thus attained below 2 atm. at 77K.

4.4. Building DSSCs

Phosphonate derivatives of the $Ru^{II}(bpy)_3$ complex were attached to γ-ZrP and placed side-by-side to different electron acceptors ranging from relatively simple viologens to more elaborated fullerene derivatives [55]. The first important observation was that the phosphorescence emission of the $Ru^{II}(bpy)_3$ complex was heavily quenched by the presence of the electron-acceptor species, indicating that the pursued electron transfer took place. More recently, flash photolysis revealed some of these materials to have outstanding properties in that the separation of charges is long lived and the initially thought inert inorganic layer resulted in semiconducting, thus driving the separated electrons and/or holes along the material [56]. Additional results showed that these powders can be arranged as solar cells with promising efficiency [57].

4.5. Molecular confinement of drugs and dyes

Confinement of drugs into molecular matrices might be a good solution to their slow release at particular locations on the body [55]. Preliminary studies show that the erythromycin family antibiotics can intercalate into ZrP phases. The measurement of the interlayer distance versus the stoichiometry of the reactants shows that mono- and bilayers of erythromycin derivatives, as an example, may be enclosed between the inorganic lamellae.

On its part, dyes are an integral part of many technologies. For example, Malachite green (MG) is used in Gram's stain which is a technique used to classify bacteria. Also, it is used for anti-fungal purposes. However, many dyes are toxic and carcinogenic and have to be handled with special care if serious environmental contamination is to be avoided. Purification of water and air resources that contain even traces of dyes is an important technological challenge and has attracted wide attention. Dye-containing coloured water is of no use but appropriately bleached solutions may still be used for washing, cooling, irrigation and cleaning purposes. Materials able to efficiently entrap dyes are thus quite desirable. **ZrP** phases serve this purpose as well (**Figure 8**) [8, 9].

These preliminary results are a good omen regarding the high prospects of using ZrP phases in the uptake of different drugs and dyes like MG or its next-of-kin methyl- and crystal violet. They also show the importance of ZrP and its surfactant composites in the field of intercalation and inclusion chemistry of bioactive organic compounds within layered inorganic and organic-inorganic materials.

4.6. Metal uptake

One measure of water quality is its 'hardness' which is defined by its Mg^{2+} and Ca^{2+} ion contents. The synthesis and design of materials towards efficiently entrapping Ca^{2+} are of great industrial importance because hard water, on being heated, precipitates calcium carbonate which clogs boilers and pipes. Moreover, Ca^{2+} reacts with soap and produces an insoluble product (soap scum). This scum is abrasive and may weaken clothes fibres.

λ-ZrP and its new ethylenediaminetetraacetic acid (EDTA)-functionalized derivative (λ-$ZrPH_2Y$) strongly complexes Ca^{2+}. The results are given in **Figure 9** [10].

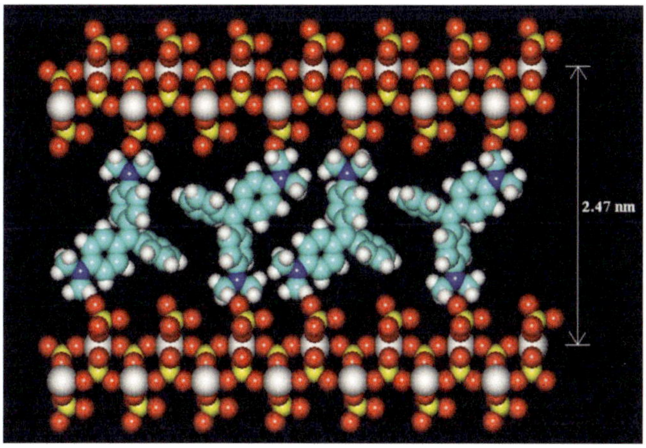

Figure 8. Possible arrangement of MG intercalated in **γ-ZrP** complying with the observed interlayer distance measured by XRD.

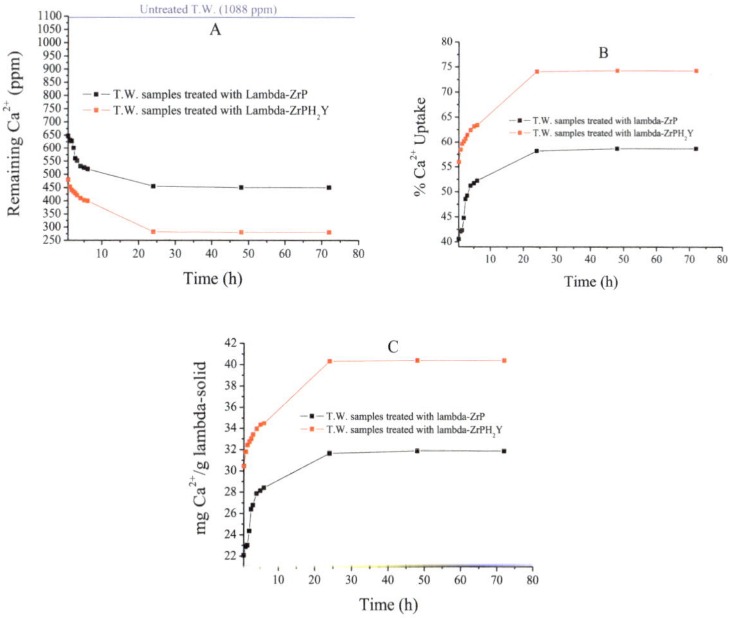

Figure 9. Kinetic curves of Ca^{2+} uptake from tap water samples (T.W.). (A) The remaining concentration of Ca^{2+} (ppm) in TW samples after treatment with λ-solids. (B) % Ca^{2+} uptake by λ-ZrP and λ-$ZrPH_2Y$. (C) Ca^{2+} uptake capacities of λ-ZrP and λ-$ZrPH_2Y$ (mg g^{-1}).

Metal uptake reached the equilibrium after ca. 24 h, where the maximum Ca^{2+} uptake is obtained (32.0 and 40.4 mg Ca^{2+}/g λ-phase, respectively), corresponding to ca. 0.8 and 1.0 mmol Ca^{2+}/g λ-phase and % Ca^{2+} uptake of 60 and 74% (**Figure 9**B), respectively.

These positive results have stimulated the testing of λ-ZrP-EDTA material for the removal of toxic metals such as Cu^{2+} and Ni^{2+} from water. The results are summarized in **Figures 10** and **11** [58].

Equilibrium was reached for Cu^{2+} and Ni^{2+} after ca. 48 h, where the maximum uptake is obtained (72.0 and 93.5 mg M^{2+}/g λ-$ZrPH_2Y$, respectively). These uptake values correspond to ca. 4.58 and 5.49 mmol M^{2+}/g λ-$ZrPH_2Y$ and % M^{2+} uptake of 58.0 and 76.6 %, respectively. This shows that λ-$ZrPH_2Y$ has higher Ni^{2+} uptake capacity as compared to Cu^{2+}.

These experiments showed that EDTA was successfully incorporated inside the interlayer gallery of λ-ZrP by means of topotactic carboxylate/Cl ligand exchange reaction. The resulting

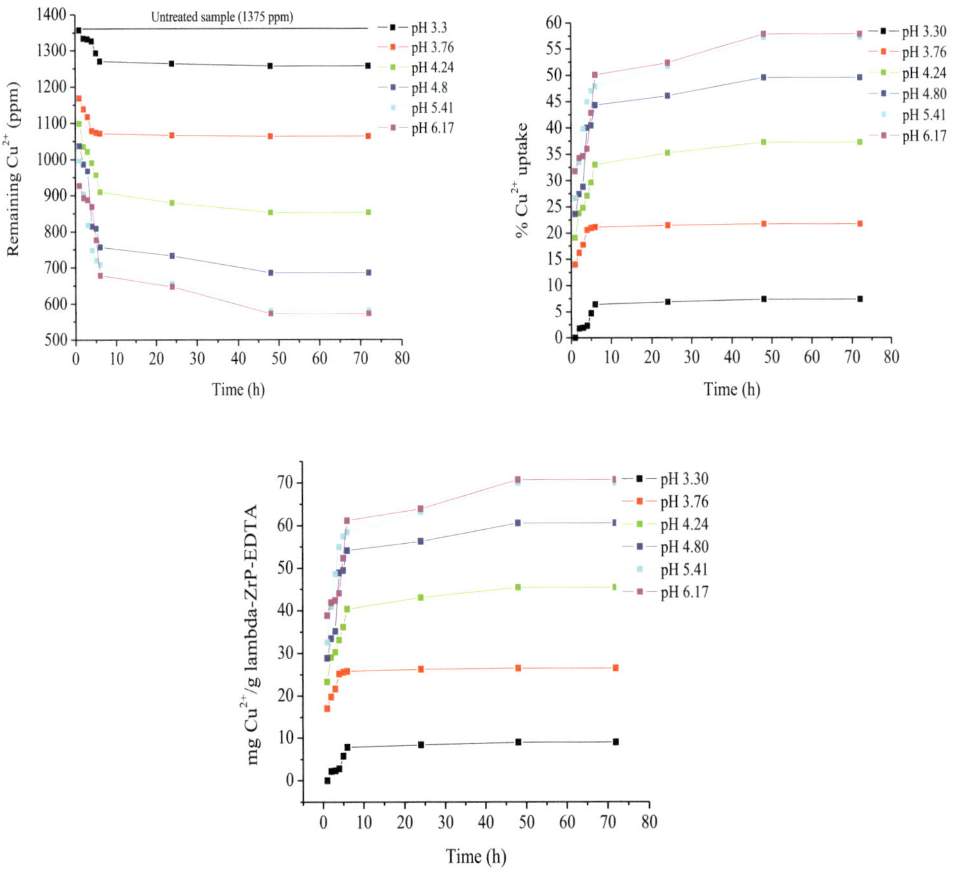

Figure 10. Kinetic curves of Cu^{2+} uptake from aqueous solutions: The remaining concentration of Cu^{2+} (ppm) in samples after treatment with λ-$ZrPH_2Y$ (*top left*). % Cu^{2+} uptake by λ-$ZrPH_2Y$ (*top right*). Cu^{2+} uptake capacities of λ-$ZrPH_2Y$ (mg g^{-1}) (*bottom*).

http://dx.doi.org/10.5772/68140

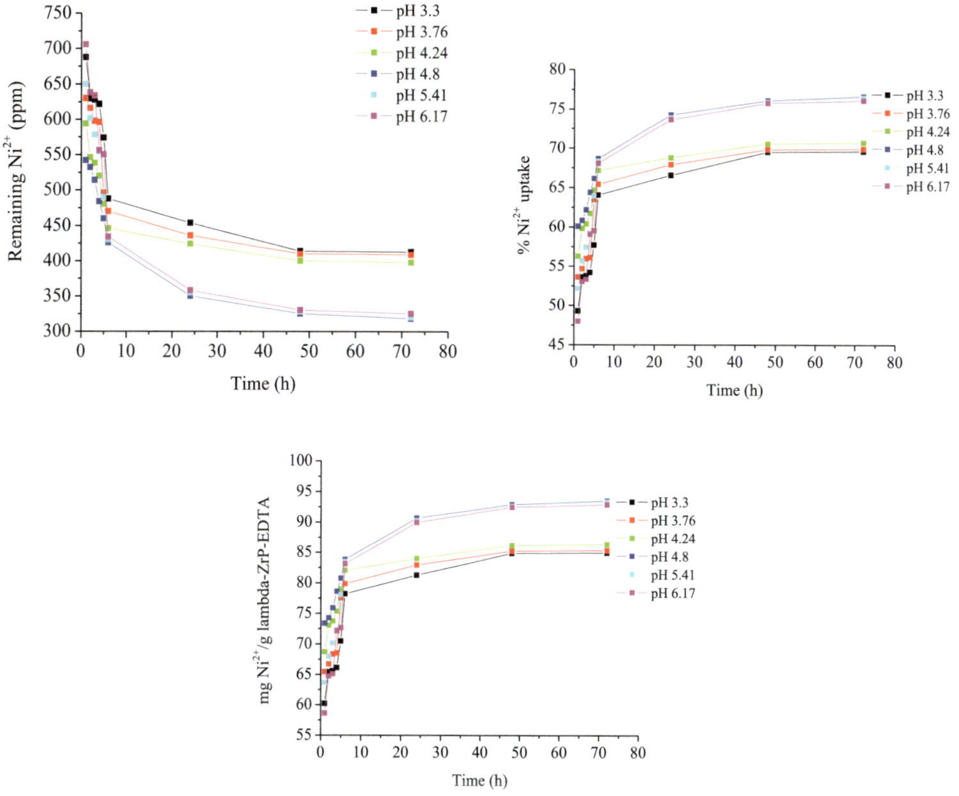

Figure 11. Kinetic curves of Ni^{2+} uptake from aqueous solutions (initial conc. 1375 ppm): The remaining concentration of Ni^{2+} (ppm) in samples after treatment with λ-$ZrPH_2Y$ (*top left*). % Ni^{2+} uptake by λ-$ZrPH_2Y$ (*top right*). Ni^{2+} uptake capacities of λ-$ZrPH_2Y$ (mg g^{-1}) (*bottom*).

composite phase, λ-$ZrPH_2Y$, exhibited high potential for the extraction of various metal ions from water samples.

5. Conclusion and prospects

Phosphoric acid chemistry has limitless features. In this article, we have gathered a collection of eloquent experimental facts where the primordial phosphoric acid, combined with Zr metal and other chemicals, led to the building of inorganic-organic 2D and 3D structures that can play a crucial role in numerous research areas. The number of rather complex composites that can be figured out from the presented rationale is inexhaustible and only limited by the imagination and chemical needs of the researcher. Moreover, the confinement of organic molecules within the ZrP frameworks would confer them new properties at the supramolecular level and give rise to many surprises driven by the high level of *serendipity* that was and will be found in this field.

Acknowledgements

This article has been written within the collaborative research work activities between the research groups of Prof. H. Ahendawi (*Al-Azhar University – Gaza, Palestine*) and Prof. E. Brunet (*Autonomous University of Madrid, Spain*). It should be mentioned that parts of this chapter are reproduced from the authors' previous publications [12].

Author details

Hussein M. H. Alhendawi[1]*, Ernesto Brunet[2] and Elena Rodríguez-Payán[2]

*Address all correspondence to: hussein.alhendawi@yahoo.com

1 Department of Chemistry, Faculty of Science, Al-Azhar University-Gaza, Gaza, Palestine

2 Department of Organic Chemistry, Faculty of Sciences, Universidad Autónoma de Madrid, Madrid, Spain

References

[1] Brunet E, Alhendawi H, Juanes O, Jiménez L, Rodríguez-Ubis JC. Luminescence of Lanthanides in Pillared Zirconium Phosphate. J. Mater. Chem. 2009; **19**:1-10. DOI: 10.1039/b817317f

[2] Brunet E, Alhendawi H, Juanes O, Rodríguez-Ubis JC. The Quest for Relationships between Conformation and Chiroptical Properties: From Solution to Solid State. J. Mex. Chem. Soc. 2009; **53(3)**:155-162. DOI: 47512154013

[3] Brunet E, de la Mata MJ, Juanes O, Alhendawi H, Cerro C, Rodríguez-Ubis JC. Solid-State Reshaping of Nanostructured Crystals: Supramolecular Chirality of Layered Materials Derived from Polyethylenoxa-Pillared Zirconium Phosphate. Tetrahedron: Asymmetry 2006; **17(3)**:347-354. DOI: 10.1016/j.tetasy.2006.01.014

[4] Brunet E, Alhendawi H, Cerro C, de la Mata MJ, Juanes O, Rodríguez-Ubis JC. Hydrogen Storage in a Highly Porous Solid Derived from γ-Zirconium Phosphate. Angew. Chem. Int. Edit. 2006; **45**:6918-6920. DOI: 10.1002/anie.200602445

[5] Brunet E, Alhendawi H, Cerro C, de la Mata MJ, Juanes O, Rodríguez-Ubis JC. Creating Libraries of Porous Materials Derived from γ-Zirconium Phosphate: Pillaring with Polyphenylethynyl Diphosphonates. Micropor. Mesopor. Mat. 2011; **138**:75-85. DOI: 10.1016/j.micromeso.2010.09.027

[6] Brunet E, Alhendawi H, Cerro C, de la Mata MJ, Juanes O, Rodríguez-Ubis JC. Easy γ-to-α Transformation of Zirconium Phosphate/Polyphenylphosphonate Salts: Porosity and Hydrogen Physisorption. Chem. Eng. J. 2010; **158**:333-344. DOI: 10.1016/j.cej.2010.01.040

[7] Brunet E, Alhendawi H, Alonso M, Cerro C, Jiménez L, Juanes O, de la Mata MJ, Salvador A, Victoria M, Rodríguez-Payán E, Rodríguez-Ubis JC. The Use of Laminar Inorganic Salts to Make Organic Molecules Display New Properties at the Supramolecular Level in the Solid State. J. Phys.: Conf. Ser. 2010; **232**:012017. DOI: 10.1088/1742-6596/232/1/012017

[8] Alhendawi H. Intercalation of Malachite Green ($[C_6H_5C(C_6H_4N(CH_3)_2)_2]Cl$) in Layered γ-Zirconium Phosphate. Effect of Cationic Surfactants. J. Mater. Chem. 2011; **21**:7748-7754. DOI: 10.1039/c0jm04147e

[9] Alhendawi H, Brunet E, Rodríguez Payán E, Juanes O, Rodríguez Ubis JC, Al-Asqalany M. Surfactant-assisted Intercalation of Crystal Violet in Layered γ-Zirconium Phosphate. Dye uptake from aqueous solutions. J. Incl. Phenom. Macrocycl. Chem. 2012; **73**:387-396. DOI: 10.1007/s10847-011-0076-6

[10] Alhendawi H, Brunet E, Juanes O, Idhair S, Hammouda H, Rodríguez Payán E, de Victoria Rodríguez M. Functionalization of lambda-zirconium phosphate with ethylene-diaminetetraacetic acid: Synthesis, characterization and applications. J. Chem. Sci. 2014; **120(6)**:1721-1727. DOI: 0.1007/s12039-014-0726-5

[11] Clearfield A. Metal phosphate chemistry. In: Karlin KD, editor. Progress in Inorganic Chemistry. New York: Wiley; 1998. p. 373-510. DOI: 978-0-471-24039-6

[12] Brunet E, de Victoria-Rodrı´guez M, García-Patrón LJ, Hindawi H, Rodríguez-Payán E, Rodríguez-Ubis JC, Juanes O. Tales from the unexpected: chemistry at the surface and interlayer space of layered organic–inorganic hybrid materials based on γ-zirconium phosphate. In: Brunet E, Colón JL, Clearfield A, editors. Tailored Organic-Inorganic Materials. Hoboken: Wiley; 2015. p. 45-82. DOI: 978-1-118-77346-8

[13] Clearfield A, Stynes JA. The Preparation of Crystalline Zirconium Phosphate and Some Observations on its Ion Exchange Behavior. J. Inorg. Nucl. Chem. 1964; **26**:117-129.

[14] Clearfield A, Blessing RH, Stynes JA. New Crystalline Phases of Zirconium Phosphate Possessing Ion-Exchange Properties. J. Inorg. Nucl. Chem. 1968; **30**:2249-2258.

[15] Poojary DM, Zhang B, Clearfield A. Synthesis and Crystal Structure of a New Layered Zirconium Phosphate Compound, $Zr(PO_4)F(OSMe_2)$. J. Chem. Soc. Dalton Trans. 1994; **16**:2453-2456. DOI: 10.1039/DT9940002453

[16] Whittingham MS, Jacobson AJ, editors. Intercalation Chemistry. New York: Academic Press; 1982. 614 p. DOI: 0124120679

[17] Tomita I, Takeo C. Synthesis of γ-Zirconium Phosphate by Fluoro-Complex Method and Intercalation Behavior of Some α-Dimines. J. Incl. Phenom. Mol. 1990; **9**:315-325.

[18] Feng Y, He W, Zhang X, Jia X, Zhao H. The Preparation of Nanoparticle Zirconium Phosphate. Mater. Lett. 2007; **61**:3258-3261. DOI: org/10.1016/j.matlet.2006.11.132

[19] Clearfield A, Costantino U. Layered Metal Phosphates and Their Intercalation Chemistry. In: Alberti G, Bein T, editors. Comprehensive Supramolecular Chemistry. New York: Pergamon; 1996. p. 107-149. DOI: 9780080912547

[20] Alhendawi H. Synthesis and Structural Characterization of Zirconium Phosphate Adipate Dimethyl Sulfoxide: A New Lambda-Type Organic-Inorganic Layered Material. J. Chem. Sc. 2014; **126(4)**: 975-979. DOI: 10.1007/s12039-014-0634-8

[21] Liu J, Boo WJ, Clearfield A, Sue HJ. Intercalation and Exfoliation: A Review on Morphology of Polymer Nanocomposites Reinforced By Inorganic Layer Structures. Mater. Manuf. Process. 2006; **20**:143-151. DOI: 10.1081/AMP-200068646

[22] Alberti G, Dionigi C, Giontella E, Murcia-Mascarós S, Vivani R. Formation of Colloidal Dispersions of Layered γ-Zirconium Phosphate in Water/Acetone Mixtures. J. Colloid Interf. Sci. 1997; **188**:27-31. DOI: org/10.1006/jcis.1996.4679

[23] Alberti G, Casciola M, Costantino U. Inorganic Ion-Exchange Pellicles Obtained By Delamination of α-Zirconium Phosphate Crystals. J. Colloid Interf. Sci. 1985; **107**: 256-263. DOI: org/10.1016/0021-9797(85)90169-9

[24] Vivani R, Alberti G, Costantino F, Nocchetti M. New Advances in Zirconium Phosphate and Phosphonate Chemistry: Structural Archetypes. Micropor. Mesopor. Mater. 2008; **107**:58-70. DOI:10.1016/j.micromeso.2007.02.029

[25] Alberti G, Vivani R, Marmottini F, Zappelli P. Microporous Solids Based on Pillared Metal(IV) Phosphates and Phosphonates. J. Porous Mater. 1998; **5**:205-220. DOI: 10.1023/A:1009678120336

[26] Alberti G. Bidimensional Structures Based on Metal Phosphates and their Applications. Sci. Tech. 1998; **80**:607-614. DOI: 29062316

[27] Alberti G, Casciola M, Costantino U, Vivani R. Layered and Pillared Metal(IV) Phosphates and Phosphonates. Adv. Mater. 1996; **8**:291-303. DOI: 10.1002/adma.19960080405

[28] Nakato T, Miyamoto N. Sol–Gel Transition of Nanosheet Colloids of Layered Niobate $K_4Nb_6O_{17}$. J. Mater. Chem. 2002; **12**:1245-1246. **DOI:** 10.1039/B202000A

[29] Miyamoto N, Nakato T. Liquid Crystalline Colloidal System Obtained by Mixing Niobate and Aluminosilicate Nanosheets: A Spectroscopic Study Using a Probe Dye. Langmuir 2003; **19**:8057. **DOI:** 10.1021/la0268449

[30] Miyamoto N, Kuroda K, Ogawa M. Exfoliation and Film Preparation of a Layered Titanate, $Na_2Ti_3O_7$, and Intercalation of Pseudoisocyanine Dye. J. Mater. Chem. 2004; **14**:165-170. DOI: 10.1039/B308800F

[31] Geng F, Ma R, Ebina Y, Yamauchi Y, Miyamoto N, Sasaki T. Gigantic Swelling of Inorganic Layered Materials: A Bridge to Molecularly Thin Two-Dimensional Nanosheets. J. Am. Chem. Soc. 2014; **136**:5491-5500. DOI: 10.1021/ja501587y

[32] Alhendawi H, Brunet E, Hammouda H, Rodríguez Payán E. Intercalation of Primary Alkylamines into Lambda-Zirconium Phosphate. Lambda-Type Materials with Extended Interlayer Separation. J. Porous. Mater. 2016; **23**:1519-1526. DOI: 10.1007/s10934-016-0212-1

[33] Caneschi A, Gatteschi D, Sangregorio C, Vaz MGF, Costantino U, Nocchetti M, Vivani R. Intercalation of a Nitronyl Nitroxide Radical into Layered Inorganic Hosts. Preparation and Physico-Chemical Characterization. Inorg. Chim. Acta. 2002; **338**:127-132. DOI: org/10.1016/S0020-1693(02)01026-5

[34] Alhendawi H, Brunet E, Juanes O, Hammouda H, Idhair S, Rodríguez Payán E, de Victoria Rodríguez M. New Soft Porous Frameworks Based On Lambda-Zirconium Phosphate And Aliphatic Dicarboxylates: Synthesis And Structural Characterization. J. Phys. Chem. Sol. 2015; **86**:95-100. DOI: org/10.1016/j.jpcs.2015.06.021

[35] Alhendawi H, Brunet E, Rodríguez Payán E, Shurrab N, Juanes O, Idhair S, Al-Asqalany M. A New Layered Zirconium Biphosphonate Framework Covalently Pillared with N,N′-Piperazinebis(Methylene) Moiety: Synthesis And Characterization. J. Porous Mater. 2013; **20(5)**:1189-1194. DOI: 10.1007/s10934-013-9702-6

[36] Alberti G, Murcia-Mascarós S, Vivan R. Preparation and Characterization Of Zirconium Phosphate Diphosphonates with the γ-Structure: A New Class Of Covalently Pillared Compounds. Mater. Chem. Phys. 1993; **35**:187-192. DOI: org/10.1016/0254-0584(93)90129

[37] Brunet E, de la Mata MJ, Alhendawi H, Cerro C, Alonso M, Juanes O, Rodriguez-Ubis JC. Engineering of Microcrystalline Solid-State Networks Using Cross-Linked γ-Zirconium Phosphate/Hypophosphite with Nonrigid Polyethylenoxadiphosphonates. Easy Access to Porously Dynamic Solids with Polar/Nonpolar Pores. Chem. Mater. 2005; **17**:1424-1433. **DOI:** 10.1021/cm048754q

[38] Alberti G, Brunet E, Chiara D, de la Mata MJ, Juanes O, Rodríguez-Ubis JC, Vivan R. Shaping Solid-State Supramolecular Cavities: Chemically Induced Accordionlike Movement of gamma-Zirconium Phosphate Containing Polyethylenoxide Pillars. Angew. Chem. Int. Ed. 1999; **22(38)**:3351-3353. DOI: 10.1002/(SICI)1521-3773(19991115)38:22<3351::AID-ANIE3351>3.0.CO;2-V

[39] Alberti G, Costantino U, Marmottini F, Vivani R, Valentini C. Ion Exchange Properties of α-Zirconium Phosphate Modified by a Process of Intercalation/De-Intercalation with Monoamines. In: Williams PA, Hundson MJ editors. Recent Development in Ion Exchange. London: Elsevier Applied Sciences; 1987. p. 233-248. DOI: 978-94-010-8044-6

[40] Wan BZ, Anthony RG, Peng GZ, Clearfield A. Characterization of Organically Pillared Zirconium Phosphates. J. Catal. 1986; **101**:19-27. DOI: 10.1016/0021-9517(86)90224-1

[41] Alberti G, Costantino U. Recent Progress in The Intercalation Chemistry of Layered α-Zirconium Phosphate And Its Derivatives, And Future Perspectives For Their Use In Catalysis. J. Mol. Catal. 1984; **27**:235-250. DOI: org/10.1016/0304-5102(84)85083-X

[42] Cao G, Garcia ME, Alcalá M, Burgess LG, Mallouk TE. Chiral Molecular Recognition in Intercalated Zirconium Phosphate. J. Am. Chem. Soc. 1992; **114**: 7574-7575. DOI: 10.1021/ja00045a046

[43] Mallouk TE, Gavin JA. Molecular Recognition in Lamellar Solids and Thin Films. Acc. Chem. Res. 1998: **31**:209-217. DOI: 10.1021/ar970038p

[44] Brunet E. Asymmetric Induction Under Confinement. Chirality 2002; **14**:135-143. DOI: 10.1002/chir.10054

[45] Vivani R, Masci S, Alberti G. Anionic Ligand Exchange on $ZrPO_4Cl$(dmso): Alkoxide and Carboxylate Derivatives. Inorg. Chem. 2004; **43**:368-374. DOI: 10.1021/ic034695r

[46] Alhendawi H. Lambda-Zirconium Phosphate Covalently Pillared with 1,4-Biphenyl-dicarboxylate: A New Rigid Mesoporous Framework. J. Incl. Phenom. Macrocycl. Chem. 2016; **85**:187-192. DOI 10.1007/s10847-016-0618-z

[47] Alhendawi H. Synthesis and Structural Characterization Of A New Chiral Porous Hybrid Organic–Inorganic Material Based On Γ-Zirconium Phosphates and L-(+)-Phosphoserine. J. Solid State Chem. 2013; **201**:24-28. DOI: org/10.1016/j.jssc.2013.02.025

[48] Alberti G, Bartocci M, Santarelli M, Vivani R. Zirconium Phosphate Chloride Dimethyl Sulfoxide, a Reactive Precursor of a Large Family of Layered Compounds. Inorg. Chem. 1997; **36**:3574-3575. DOI: 10.1021/ic9704389

[49] Alberti G, Masci S, Vivani R. Layered Zirconium Phosphate Chloride Dimethyl Sulfoxide as a Two-Dimensional Exchanger of Anionic Ligands. Part I. Substitution of Chloride with Inorganic Monodentate Ligands. Inorg. Chem. 2002; **41**:1913-1919. DOI: 10.1021/ic010643v

[50] Brunet E, Huelva M, Rodriguez-Ubis JC. Covalent Bonding of Aza-18-Crown-6 to γ-Zirconium Phosphate. A New Layered Ion-Exchanger with Potential Recognition Capabilities .Tetrahedron. Lett. 1994; **35**:8697-8700. DOI: org/10.1016/S0040-4039(00)78475-9

[51] Brewster JH. A Useful Model of Optical Activity. I. Open Chain Compounds. J. Am. Chem. Soc. 1959; **81**:5475-5483. DOI: 10.1021/ja01529a056

[52] Eliel EL, Brunet E. Quantitative Relationship Between Optical Rotation And Conformation J. Org. Chem. 1991; **56**:1668-1670. DOI: 10.1021/jo00004a059

[53] Brunet E, Juanes O, Rodriguez-Ubis JC. Supramolecularly Organized Lanthanide Complexes for Efficient Metal Excitation and Luminescence as Sensors in Organic and Biological Applications. Curr. Chem. Biol. 2007; **1**:11-39. DOI: 10.2174/2212796810701010011

[54] Brunet E, Cerro C, Juanes O, Rodriguez-Ubis JC, Clearfield A. Hydrogen Storage in Highly Microporous Solids Derived From Aluminium Biphenyldiphosphonate. J. Mater. Sci. 2008; **43(3)**:1155-1158. DOI 10.1007/s10853-007-2377-0

[55] Salonen J, Kaukonen AM, Hirvonen J, Lehto VP. Mesoporous Silicon in Drug Delivery Applications. J. Pharm. Sci. 2008; **97(2)**:632-653. DOI: 10.1002/jps.20999

[56] Brunet E, Alonso M, Quintana MC, Atienzar P, Juanes O, Rodriguez-Ubis JC, Garcia H. Laser Flash-Photolysis Study of Organic-Inorganic Materials Derived from Zirconium Phosphates/Phosphonates of Ru(bpy)3 and C60 as Electron Donor-Acceptor Pairs. J. Phys. Chem. C 2008; **112(15)**:5699-5702. DOI: 10.1021/jp800026r

[57] Teruel L, Alonso M, Quintana MC, Salvador A, Juanes O, Rodriguez-Ubis JC, Brunet E, Garcia H. Photovoltaic Activity of Layered Zirconium Phosphates Containing Covalently Grafted Ruthenium Tris(Bipyridyl) and Diquat Phosphonates as Electron Donor/Acceptor Sites. Phys. Chem. Chem. Phys. 2009; **11(16)**:2922-2927. DOI: 10.1039/B816698F

[58] Alhendawi H, Brunet E, Hammouda H, Abu Tuha G, Rodríguez Payán E. To be published.

Chapter 7

Purification of Phosphoric Acid by Liquid-Liquid Equilibrium

Khatereh Bahrpaima

Additional information is available at the end of the chapter

http://dx.doi.org/10.5772/67926

Abstract

Various ternary and quaternary liquid-liquid phase equilibrium data for water + phosphoric acid + solvent(s) have been reported. Salting-out, solvent, and temperature effects on the binodal curve and the tie lines have been highlighted and the capability of solvents with different functional groups to extract phosphoric acid from water has been compared. Studying of influence of magnetic, electromagnetic, and ultrasonic fields on the separation factors and distribution coefficients of aqueous phosphoric acid mixtures has been proposed. Moreover, a summary of the optimized binary interaction values, which resulted from non-random two-liquid (NRTL) and universal quasi-chemical (UNIQUAC) thermodynamic models using genetic algorithm (GA), bee algorithm (BA), and simulated annealing (SA), has been presented. Group method of data handling (GMDH) and linear solvation energy relationship (LSER) methods for the correlation of experimental liquid-liquid equilibrium (LLE) data have been used.

Keywords: phosphoric acid, LLE, optimization algorithms, GMDH, LSER

1. Introduction

The phosphoric acid due to its potential applications in various industries attracted a great deal of attention in the area of petrochemical operations, pharmaceutical productions, food industry, detergents, insecticides, and agricultural fertilizers. Generally, there are two basic methods in commercial use for the production of phosphoric acid—thermal and wet processes. The thermal process produces a pure acid with huge energy consumption, whereas the wet process is economic and practiced everywhere in the world [1]. Nevertheless, phosphoric acid produced from both methods contains a variety of impurities, which could affect quantity and the quality of the product. Therefore, for efficient removal and decreasing the impurities to below acceptable regulatory levels, not only suitable studies but also using and developing new different techniques and technologies are required. For this purpose, some of the common selective methods such as liquid-liquid extraction, solid-liquid extraction, crystallization, concentration, addition of additives, and so on, have been investigated [2–4].

Liquid-liquid extraction or solvent extraction is now one of the most important and widely employed separation techniques in the modern process industries. It is well known due to various practical characteristics, including simplicity, rapid method development, and reasonable selectivity. Moreover, it can be used for several purposes, for instance, to separate the systems with similar boiling points, high boiling points, and temperature-sensitive components [2].

Success of liquid-liquid extraction and accurate understanding of the performance of that is measured through the liquid-liquid equilibrium (LLE) data. LLE is described with the separation of components of a feed solution, containing the carrier and extract components, using an additional liquid solvent, as the mixture of feed and solvent—if enough solvent is added—forms two immiscible liquid phases. LLE is based on the differences in solubility and equilibrium distribution of these components between the two produced immiscible—or partially miscible—phases. In other words, it depends on the mass transfer of the component to be extracted from the carrier to the solvent. For having an effective extraction, the extract component should more preferably dissolve in the solvent. After settling the two phases, the raffinate and the extract phases are formed. The raffinate phase contains mostly the carrier, as well as a residue of the extracted component and solved solvent, whereas the extract phase consists of the main the solvent, a part of the extracted component, and solved carrier. Solvent extraction is a separation process aiming to purify the feed or to recover one or more components from it [5]. Nowadays, enormous amounts of acid, specifically high-purity phosphoric acid, can be commercially produced through liquid-liquid equilibrium process.

2. Phase diagram of multicomponent liquid mixtures: measurement and visualization

Only systems with at least three components and a miscibility gap can be used for extractions. Phase behavior of such systems at a constant temperature and pressure is conveniently represented on an equilateral triangular diagram, which its corners indicate the pure components, binary compositions are along the edges, and ternary mixtures are located inside the triangle. A common phase diagram is type I system and shown in **Figure 1**, where a pair of components propylene carbonate and water are partially miscible, and liquid phosphoric acid dissolves completely in propylene carbonate or water [6]. In this figure, the boundaries that each one of them separates the single-phase region from the two-phase region are named binodal (solubility) curves. The two-phase region is included inside below the curved envelope. The binodal curve results are determined by cloud-point titration method [7]. According to this method, a binary mixture of known composition was titrated with the third component at each mixture using non-sealed glass vessels. The transition point is taken as the appearance/disappearance of turbidity in the sample and it is defined as a cloud point. The mixture temperature is regulated by a thermostatic thermometer with an accuracy of ± 0.1 K. Every point on the binodal curve also has another corresponding point on the binodal curve, as these two points represent the phase equilibrium. The lines between these two equilibrium points

http://dx.doi.org/10.5772/67926

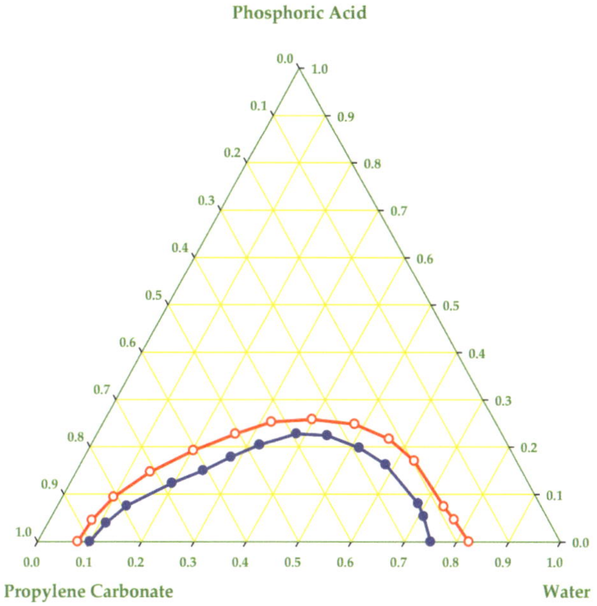

Figure 1. Binodal curves for (water + phosphoric acid + propylene carbonate) mixture at two different temperatures and atmospheric pressure: (-○-) $T = 298.2$ K, and (-●-) $T = 318.2$ K [6].

are called tie lines. The tie lines are usually not parallel and their slopes can increase and decrease drastically. Tie line experiments were carried out in a jacketed 150 ml glass cell. The biphasic mixture with known compositions was placed in the extraction cell and was vigorously agitated by a magnetic stirrer for 4 h, and then left to settle for minimum 4 h for phase

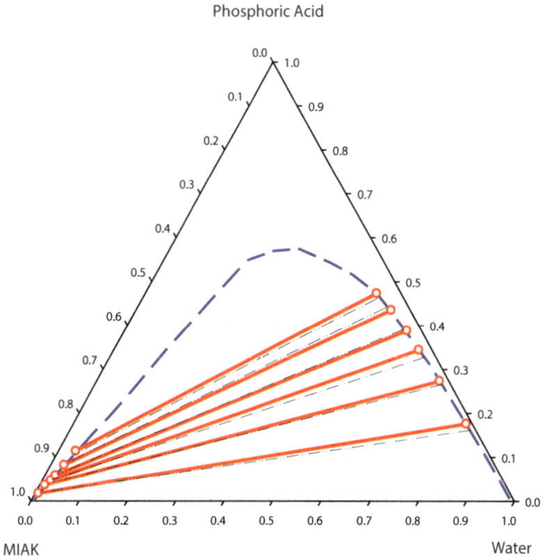

Figure 2. Tie lines of the system water + phosphoric acid + MIAK at two different temperatures and atmospheric pressure: (-○-) $T = 308.2$ K, and (–⋯–), $T = 318.2$ K [9].

separation. After separation, samples of both phases were transparent and were carefully weighed and analyzed to determine their compositions. Then, the concentrations of the acid in both the phases were obtained by potentiometric NaOH titration [6]. Besides, the water content of the organic phase was measured by the Karl-Fisher method [6] and the water contained in the aqueous layer was determined using refractive index measurement method [8]. The obtained tie line data for (water + phosphoric acid + methyl isoamyl ketone (MIAK)) [9] and (water + phosphoric acid + dichloromethane (DCM)) [10] ternary systems are presented in **Figure 2** and **Table 1**, respectively. As either the temperature or pressure is varied, the location of the binodal curve and slopes of the tie lines may change (see **Figures 1** and **2**).

Aqueous phase (raffinate) mass percent			**Organic phase (extract) mass percent**			D_1	D_2	S
w_{11}	w_{21}	w_{31}	w_{13}	w_{23}	w_{33}			
70.48	15.00	14.52	6.14	1.40	92.46	0.087	0.093	1.071
66.82	16.39	16.79	6.49	1.67	91.84	0.097	0.101	1.049
58.19	19.78	22.03	6.60	2.01	91.39	0.113	0.102	0.896
50.89	23.28	25.83	6.78	2.92	90.30	0.133	0.125	0.941
45.74	24.33	29.93	6.80	3.31	89.89	0.149	0.136	0.915

*GMDH is used to predict the tie lines.

Table 1. Calculated tie line data, distribution coefficient, and separation factor for the ternary system of water (1) + phosphoric acid (2) + DCM (3), according to the GMDH* results [10].

3. Separation factor and distribution coefficients

In order to evaluate the extracting capability of the solvent for the separation of components from feed solutions with liquid-liquid extraction, the separation factor, S, is calculated. The separation factor is defined as the ratio of distribution coefficients of extract component, 2, to carrier one, 1, $S = D_2/D_1$. Additionally, distribution coefficient is the ratio of concentrations of a component in a mixture of two immiscible phases at equilibrium. This ratio is therefore a measure of the difference in the component solubilities in these two phases (see **Table 1**). Depending on the system, the distribution coefficient and separation factor can be a function of temperature, the concentration of chemical species in the system, and a large number of other parameters. **Figure 3** gives the distribution coefficient as a function of the mass percent of phosphoric acid in aqueous phase for water + phosphoric acid + mixed-solvent (dichloromethane + 1,2-dichloroethane (DCE)) system at different ratios of mixed solvent [11]. Generally, if the separation factor is greater than one ($S > 1$) for the systems investigated, it means that the component can be extracted by the proposed solvent—for example, according to the reported S for studied system in **Table 1**, DCM can be considered as a possible candidate for the recovery of the aqueous phosphoric acid solutions.

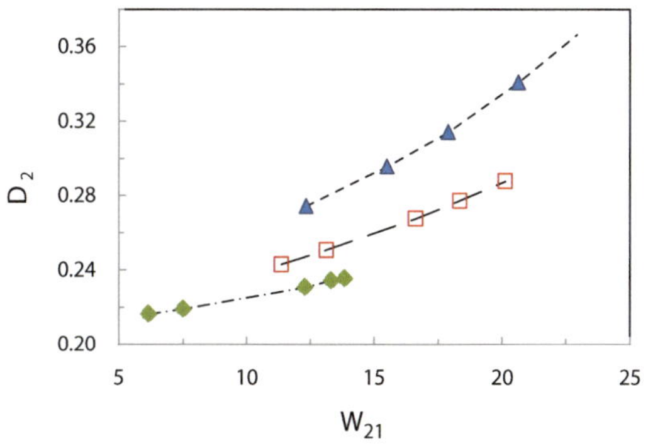

Figure 3. Distribution coefficients D2 of phosphoric acid as a function of the mass percent W_{21} of acid in aqueous phase for NRTL LLE data of {water + phosphoric acid + mixed-solvent (DCM + DCE)} system at temperature T = 298.15 K: (25% DCM + 75% DCE); ◆, (50% DCM + 50% DCE); □, (75% DCM + 25% DCE); ▲ [11].

In recent years, significant investigations on LLE measurements and the extraction of phosphoric acid from aqueous solutions have been carried out, while many solvents have mainly been tested, in order to understand and provide further information for accurate interpretation of phase equilibria of the different multicomponent mixtures with phosphoric acid [12–27]. As a detailed evaluation of extraction abilities of the different solvents reported in the literature, including alcohols, esters, ethers, ketones, and hydrocarbons to extract phosphoric acid from aqueous solutions, a comparative study is listed in **Table 2**. **Figure 3** also highlights the definitive effects of different chemical structures present in solvents on recovering of the acid. As it is drawn, results show the following order of effectiveness of solvent for systems containing phosphoric acid: methylcyclohexane > ethyl benzene > isoamyl acetate > n-pentane > 1-octanol > bis(2-methylpropyl) ether. From the obtained experimental results, it can be concluded that methylcyclohexane is the most appropriate solvent for the separation of a mixture of water and phosphoric acid. It is also apparent from **Figure 4** that bis(2-methylpropyl) ether is a less favorable solvating agent for phosphoric acid.

In addition, the presence of dissolved salt changes the phase equilibrium behavior of a mixture significantly [28–31]. Recently, Govindarajan and Sabarathinam [32], Mohsen-Nia et al. [33], and Santos et al. [34] have investigated effects of some inorganic salts such as Na_2SO_4, $ZnSO_4$, $(NH_4)_2SO_4$, NaCl, KCl, KBr, and $NaNO_3$ on the ternary liquid-liquid equilibria data. Similarly, by considering the importance of the salting effect on the extraction of phosphoric acid from aqueous mixtures, it is worthwhile to study LLE of mixtures of (water + phosphoric acid + solvent + salt). In particular, as can be seen in **Table 2**, maximum values of separation factor for (water + phosphoric acid + DCM + 10 wt% NaCl and 10 wt% $CaCl_2$) solutions are improved compared with those in the absence of added salts [26].

Although solvents with high separation are considered as potential candidates to carry out the extraction, due to the economic, environmental, and technical factors, the suitable solvent at optimum conditions is proposed to be chosen for the extraction of phosphoric acid from aqueous mixtures. Meanwhile, it seems that the measurement of the (liquid + liquid) equilibrium data of

System	Solvent type	*T*/K	*S*	Ref.
Water + phosphoric acid + 1-butanol	Alcohol	308.2	1.65	[12]
Water + phosphoric acid + 2-methyl-2-butanol	Alcohol	298.2	1.82	[13]
Water + phosphoric acid + isoamy1 alcohol	Alcohol	308.2	1.82	[14]
Water + phosphoric acid + 2-ethyl-1-hexanol	Alcohol	298.2	1.46	[15]
Water + phosphoric acid + 1-octanol	Alcohol	298.2	6.35	[16]
Water + phosphoric acid + 1-nanonol	Alcohol	298.2	3.54	[17]
Water + phosphoric acid + 1-decanol	Alcohol	303.2	3.80	[18]
Water + phosphoric acid + 1-dodecanol	Alcohol	303.2	4.30	[18]
Water + phosphoric acid + cyclohexanol	Alcohol, cyclic alcohol	308.2	1.57	[14]
Water + phosphoric acid + 1-dodecanethiol	Thiol, aliphatic thiol	303.2	18.20	[18]
Water + phosphoric acid + butyl acetate	Ester	308.2	5.78	[12]
Water + phosphoric acid + isobutyl acetate	Ester	298.2	7.46	[13]
Water + phosphoric acid + isoamyl acetate	Ester	298.2	19.70	[9]
Water + phosphoric acid + hexyl acetate	Ester	298.2	7.00	[19]
Water + phosphoric acid + cyclohexyl acetate	Ester	298.2	6.20	[19]
Water + phosphoric acid + propylene carbonate	Ester	298.2	2.82	[6]
Water + phosphoric acid + tributyl phosphate	Ester	298.2	3.40	[20]
Water + phosphoric acid + diisopropyl ether	Ether	293.2	1.89	[21]
Water + phosphoric acid + bis(2-methylpropyl) ether	Ether	298.2	3.76	[22]
Water + phosphoric acid + n-pentane	Hydrocarbon	308.2	15.08	[23]
Water + phosphoric acid + 2-methylpentane	Hydrocarbon	308.2	3.36	[23]
Water + phosphoric acid + n-hexane	Hydrocarbon	308.2	4.60	[23]
Water + phosphoric acid + cyclohexane	Hydrocarbon, cycloalkane	308.2	15.30	[24]
Water + phosphoric acid + methyl cyclohexane	Hydrocarbon, cycloalkane	308.2	43.10	[24]
Water + phosphoric acid + toluene	Hydrocarbon, aromatic hydrocarbon	308.2	26.50	[24]
Water + phosphoric acid + ethyl benzene	Hydrocarbon, aromatic hydrocarbon	298.2	32.70	[25]
Water + phosphoric acid + isopropyl benzene	Hydrocarbon, aromatic hydrocarbon	298.2	23.70	[25]
Water + phosphoric acid + dichloromethane (DCM)	Halogenated aliphatic hydrocarbon	298.2	1.04	[26]
Water + phosphoric acid + DCM + 10 wt% NaCl	Halogenated aliphatic hydrocarbon + salt	298.2	2.20	[26]
Water + phosphoric acid + DCM + 10 wt% $CaCl_2$	Halogenated aliphatic hydrocarbon + salt	298.2	1.39	[26]
Water + phosphoric acid + 1,2-dichloroethane (DCE)	Halogenated aliphatic hydrocarbon	298.2	1.10	[26]
Water + phosphoric acid + DCE + 10 wt% NaCl	Halogenated aliphatic hydrocarbon + salt	298.2	2.35	[26]
Water + phosphoric acid + DCE + 10 wt% $CaCl_2$	Halogenated aliphatic hydrocarbon + salt	298.2	2.13	[26]
Water + phosphoric acid + methyl isobutyl ketone	Ketone	308.2	1.26	[14]
Water + phosphoric acid + methyl isoamyl ketone	Ketone	298.2	11.00	[9]
Water + phosphoric acid + methyl ethyl ketone	Ketone	308.2	0.91	[27]

Table 2. The maximum values of separation factors for the (water + phosphoric acid + solvent) ternary systems.

http://dx.doi.org/10.5772/67926

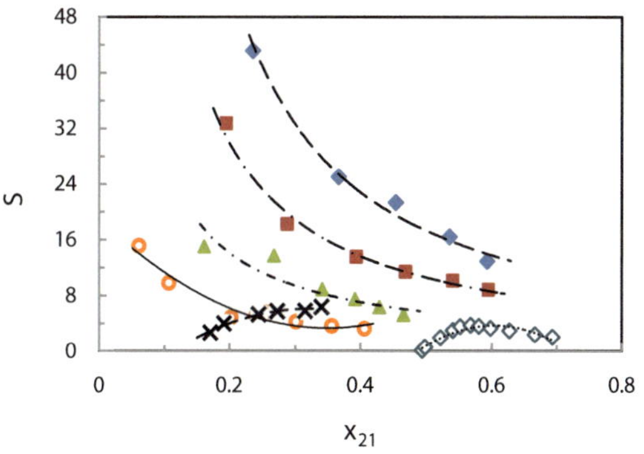

Figure 4. The effects of solvents with various functional groups on separation factor S for {water + phosphoric acid + solvent} system at mentioned temperature: (methylcyclohexane, 308.2 K [24]), ◆; (ethyl benzene, 298.2 K [25]), ■; (isoamyl acetate, 308.2 K [9]), ▲; (n-pantane, 308.2 K [23]), O; (1-octanol, 298.2 K [16]), ×; bis(2-methylpropyl) ether, 298.2 K [22]), ◇.

the mixtures with a mixed solvent has become the turning point topic of the chemistry investigations in recent years [11, 35–42]. As an illustration, **Figure 3** compares distribution coefficients of phosphoric acid for three quaternary systems of (water + phosphoric acid + mixed solvent (25% DCM + 75% DCE) or (50% DCM + 50% DCE) or (75% DCM + 25% DCE)) [11]. The results confirm the different effects of the dichloromethane as co-solvent on 1,2-dichloroethane for the extraction of phosphoric acid from aqueous mixtures. The comparison between the experimental separation factors obtained for the mixed solvent (DCM + DCE) with various ratios also indicates that mixed solvent (25% DCM, 75% DCE) has a higher separation factor than the other mixed solvents at $T = 298.15$ K, which means that the extraction of phosphoric acid by mixed solvent (25% DCM, 75% DCE) is more suitable.

Besides, it is anticipated that the magnetic, electromagnetic, and ultrasonic fields as external factors can affect LLE data by influencing solvents properties such as polarity and permittivity, especially for polar solvents. Therefore, the application of them to the phase separation in the solvent extraction process has been considered from the technical and economical viewpoints. For studying this purpose, the effect of the applied different fields on phase behavior of several ternary systems has been obtained and reported [43–48]. There are a number of effective potentials for future research to evaluate the effects of the magnetic, electromagnetic, and ultrasonic fields on the extraction of phosphoric acid from aqueous mixtures.

The linear solvation energy relationship (LSER) models [49] such as Kamlet (Eq. (1)) and Katritzky (Eq. (2)) equations [50, 51] were used to correlate separation factor for several ternary LLE systems with phosphoric acid, reporting in references [16, 19, 24]. Kamlet LSER model with the solvatochromic parameters is defined according to the following equation:

$$\log S = \log S_0 + s(x\pi^* + dx\delta) + ax\alpha + bx\beta \tag{1}$$

where π^* is a measure of solvent dipolarity/polarizability, δ is a discontinuous polarizability correction term, α is a measure of the solvent hydrogen-bond donor acidity, and β is a measure of the hydrogen-bond acceptor basicity. The coefficients S_0, s, d, a, and b include the properties of solute and are derived from regression. The values of Kamlet-Taft solvatochromic parameters (α, β, and π^*) of the solvents are obtained from literature [49].

Katritzky [51] introduced a multiparameter polarity scale with combination of Riechardt's solvent polarity, E_T^N, the dielectric constant, ε, and the index of refraction, n, functions. The Katritzky equation (Eq. (2)) allows one to estimate independent description of solvent dipolarity, polarizability, and specific interactions (such as hydrogen bonding, π-π interaction). The coefficients, a, b, c, and d measure the relative susceptibilities of log (S) to the indicated solvent parameters and are regressed using experimental data. E_T^N, ε, and n values of solvents are also present in literature [49]. A modified form of the Katritzky equation may be used, consisting of the addition of a dimensionless term as T/298.2 that characterizes temperature effect

$$\log S = \log S_0 + a\,.\,E_T^N + b\,.\left(\frac{\varepsilon - 1}{2\varepsilon + 1}\right) + c\,.\left(\frac{n^2 - 1}{2n^2 + 1}\right) + d\,.\left(\frac{T}{298.2}\right) \tag{2}$$

The LSER model values showed a good regression of the experimental data for all investigated systems. The temperature susceptibility coefficient, d, shows that all the systems in references [24] are temperature sensitive.

4. LLE correlation

Various thermodynamic models with different abilities can be used to accurately describe the multicomponent LLE systems. Aided by these activity coefficient equations and a parameter fit, it is possible to reproduce the measurement and control the data. The non-random two-liquid (NRTL) [52] and the universal quasi-chemical (UNIQUAC) [53] methods have been successfully applied for the correlation of many ternary and quaternary liquid-liquid phase equilibrium solutions [12, 16–19, 24–26], while a group contribution approach (UNIFAC) [54] may be utilized to predict the LLE mixtures.

4.1. NRTL equation

The basic idea in NRTL equation follows from the concept of local composition which was used by Renon and Prausnitz [52]. Renon's equation is applicable to partially miscible liquid mixtures, and for the binary mixtures the excess Gibbs energy is

$$\frac{G^E}{\mathrm{RT}} = x_1 x_2 \left(\frac{\tau_{21} G_{21}}{x_1 + x_2 G_{21}} + \frac{\tau_{12} G_{12}}{x_2 + x_1 G_{12}}\right) \tag{3}$$

where

$$G_{12} = \exp\left(-\alpha_{12}\tau_{12}\right);\; G_{21} = \exp\left(-\alpha_{21}\tau_{21}\right) \tag{4}$$

$$\tau_{12} = \frac{g_{12} - g_{22}}{RT};\; \tau_{21} = \frac{g_{21} - g_{11}}{RT} \tag{5}$$

The significance of g_{ij} is an energy parameter characteristic of the *i-j* interaction. Parameter α_{12} is related to the nonrandomness in the mixture; when α_{12} is zero, the mixture is completely random and Eq. (1) reduces to the two-suffix Margules equation [55]. The NRTL equation contains three parameters, but reduction of experimental data for a large number of binary systems indicates that α_{12} varies from about 0.20 to 0.47; when experimental data are scarce, the value of α_{12} can often be set arbitrarily; a typical choice is $\alpha_{12} = 0.3$. From Eq. (3), the activity coefficients are

$$\ln\gamma_1 = x_2^2\left[\tau_{21}\left(\frac{G_{21}}{x_1 + x_2G_{21}}\right)^2 + \frac{\tau_{12}G_{12}}{(x_2 + x_1G_{12})^2}\right] \tag{6}$$

$$\ln\gamma_2 = x_1^2\left[\tau_{12}\left(\frac{G_{12}}{x_2 + x_1G_{12}}\right)^2 + \frac{\tau_{21}G_{21}}{(x_1 + x_2G_{21})^2}\right] \tag{7}$$

For strongly nonideal mixtures and especially for partially immiscible systems [56], the NRTL equation often provides a good representation of experimental data if care is exercised in data reduction to obtain the adjustable parameters.

The equations discussed are readily to as many components as desired without any additional assumptions and without introducing any constants other than those obtained from binary data. For a solution of *m* components, the NRTL equation is

$$\frac{G^E}{RT} = \sum_{i=1}^{m} x_i \frac{\sum_{j=1}^{m} \tau_{jl}G_{ji}x_j}{\sum_{l=1}^{m} G_{li}x_1} \tag{8}$$

where

$$\tau_{ij} = \frac{g_{ij} - g_{jj}}{\mathrm{RT}}, \qquad \tau_{ji} = \frac{g_{ji} - g_{ii}}{\mathrm{RT}} \tag{9}$$

$$G_{ji} = \exp(-\alpha_{ji}\tau_{ji}) \quad (\alpha_{ji} = \alpha_{ij}) \tag{10}$$

The activity coefficient for any component *i* is given by

$$\ln\gamma_i = \frac{\sum_{j=1}^{m}\tau_{ji}G_{ji}x_j}{\sum_{l=1}^{m}G_{li}x_l} + \sum_{j=1}^{m}\left[\frac{x_jG_{ij}}{\sum_{l=1}^{m}G_{ij}x_l}\left(\tau_{ij} - \frac{\sum_{r=1}^{m}x_r\tau_{rj}G_{rj}}{\sum_{l=1}^{m}G_{lj}x_l}\right)\right] \tag{11}$$

Eqs. (8) and (11) contain only parameters obtained from binary data. The experimental tie line LLE data of multicomponent mixtures can be correlated using the NRTL model within well-known simulation software such as Aspen Plus. The quality of the correlation is measured by the root-mean-square deviation (RMSD) [16]. Many NRTL LLE values for the ternary and quaternary systems containing water, phosphoric acid, and solvent have appeared in the

i-j	A_{ij}(K) *	A_{ji}(K)
1–2	7.01	62.86
1–3	996.58	334.13
2–3	230.64	270.65

*g_{ij} (J mol^{-1}) = A_{ij} (K)/RT.

Table 3. Correlated LLE results from the NRTL (α_{12} = 0.3) model using SA; the corresponding binary interaction parameters, A_{ij} and A_{ji}, for ternary system water (1) + phosphoric acid (2) + DCE (3) + 10 wt% $CaCl_2$ at temperature T = 298.15 K.

i-j	A_{ij}(K)	A_{ji}(K)
1–2	2492.75	−419.50
1–3	2303.05	1212.55
1–4	7.56	−3652.94
2–3	2466.50	4329.12
2–4	−2316.04	−2817.74
3–4	−854.74	−3413.39

Table 4. NRTL (α_{12} = 0.3) interaction parameters, A_{ij} and A_{ji}, using BA for quaternary LLE system water (1) + phosphoric acid (2) + (50% DCM (3) + 50% DCE (4)) at temperature T = 298.15 K.

available literature [11, 16, 17]. For example, in **Figure 3**, the calculated NRTL data have been used to plot and show the variation of distribution coefficients of acid for (water + phosphoric acid + mixed-solvent (DCM + DCE)) system at temperature T = 298.15 K. Furthermore, several optimization algorithms such as genetic algorithm (GA) [57], bee algorithm (BA) [58], and simulated annealing (SA) [59] can be applied to predict the binary interaction parameters. As can be seen in **Table 3**, the SA and NRTL models have been used to estimate the optimized binary interaction parameters for an aqueous ternary system containing phosphoric acid [60]. **Table 4** reports 12 NRTL binary interaction parameters obtained using BA for the quaternary aqueous mixture including phosphoric acid and (50% DCM + 50% DCE) [61].

4.2. UNIQUAC equation

A critical examination of the derivation of the NRTL equation shows that this equation is more suitable for H^E than G^E [53]. Abrams derived an equation that, in a sense, extends the quasi-chemical theory of Guggenhiem [56] for nonrandom mixtures to solutions containing molecules of different size. This extension was therefore called the universal quasi-chemical theory or, in short, UNIQUAC. The UNIQUAC equation for G^E consists of two parts: a combinatorial part that attempts to describe the dominant entropic contribution and a residual part that is due primarily to intermolecular forces that are responsible for the enthalpy of mixing. The combinatorial part is determined only by the composition and by the sizes and shapes of the molecules; it requires only pure-component data. The residual part, however, depends also on

http://dx.doi.org/10.5772/67926

intermolecular forces; the two adjustable binary parameters, therefore, appear only in the residual part. The UNIQUAC equation is

$$\frac{G^E}{RT} = \left(\frac{G^E}{RT}\right)_{\text{combinatorial}} + \left(\frac{G^E}{RT}\right)_{\text{residual}} \tag{12}$$

For a binary mixture,

$$\left(\frac{G^E}{RT}\right)_{\text{combinatorial}} = x_1 \ln\frac{\Phi_1}{x_1} + x_2 \ln\frac{\Phi_2}{x_2} + \frac{z}{2}\left(x_1 q_1 \ln\frac{\theta_1}{\Phi_1} + x_2 q_2 \ln\frac{\theta_2}{\Phi_2}\right) \tag{13}$$

$$\left(\frac{G^E}{RT}\right)_{\text{residual}} = -x_1 q'_1 \ln\left(\theta'_1 + \theta'_2 \tau_{21}\right) - x_2 q'_2 \ln\left(\theta'_2 + \theta'_1 \tau_{12}\right) \tag{14}$$

where the coordination number z is set equal to 10. Segment fraction, Φ, and area fraction, θ and θ', are given by

$$\Phi_1 = \frac{x_1 r_1}{x_1 r_1 + x_2 r_2}, \qquad \Phi_2 = \frac{x_2 r_2}{x_1 r_1 + x_2 r_2} \tag{15}$$

$$\theta_1 = \frac{x_1 q_1}{x_1 q_1 + x_2 q_2}, \qquad \theta_2 = \frac{x_2 q_2}{x_1 q_1 + x_2 q_2} \tag{16}$$

$$\theta'_1 = \frac{x_1 q'_1}{x_1 q'_1 + x_2 q'_2}, \qquad \theta'_2 = \frac{x_2 q'_2}{x_1 q'_1 + x_2 q'_2} \tag{17}$$

where

$$l_j = \frac{z}{2}(r_j - q_j) - (r_j - 1) \tag{18}$$

Parameters r, q, and q' are pure component molecular structure constants depending on molecular size and external surface areas. In the original formulation, $q = q'$. To obtain better agreement for systems containing water or lower alcohols, q' values for water and alcohols were adjusted empirically by Anderson [62] to give an optimum fit to a variety of systems containing these components. For alcohols, the surface of interaction q' is smaller than the geometric external surface q, suggesting that intermolecular attraction is dominated by the OH group (hydrogen bonding). For fluids other than water or lower alcohols, $q = q'$.

For each binary mixture, there are two adjustable parameters, τ_{12} and τ_{21}. These, in turn, are given in terms of characteristic energies Δu_{12} and Δu_{21}, by

$$\tau_{12} = \exp\left(-\frac{\Delta u_{12}}{RT}\right) \equiv \exp\left(-\frac{a_{12}}{T}\right) \tag{19}$$

$$\tau_{21} = \exp\left(-\frac{\Delta u_{21}}{RT}\right) \equiv \exp\left(-\frac{a_{21}}{T}\right) \tag{20}$$

For many cases, Eqs. (19) and (20) give the primary effect of temperature on τ_{12} and τ_{21}. Characteristic energies Δu_{12} and Δu_{21} are often only weakly dependent on temperature.

$$\ln\gamma_1 = \ln\frac{\Phi_1}{x_1} + \frac{z}{2}\, q_1 ln\frac{\theta_1}{\Phi_1} + \Phi_1\left(l_1 - \frac{r_1}{r_2} l_2\right) - q_1' \ln(\theta_1' + \theta_2'\tau_{21}) + \theta_2' q_1'\left(\frac{\tau_{21}}{\theta_1' + \theta_2'\tau_{21}} - \frac{\tau_{12}}{\theta_2' + \theta_1'\tau_{12}}\right) \tag{21}$$

$$\ln\gamma_2 = \ln\frac{\Phi_2}{x_2} + \frac{z}{2}\, q_2 ln\frac{\theta_2}{\Phi_2} + \Phi_2\left(l_2 - \frac{r_2}{r_1} l_1\right) - q_2' \ln(\theta_2' + \theta_1'\tau_{12}) + \theta_1' q_2'\left(\frac{\tau_{12}}{\theta_2' + \theta_1'\tau_{12}} - \frac{\tau_{21}}{\theta_1' + \theta_2'\tau_{21}}\right) \tag{22}$$

For a multicomponent system, the UNIQUAC equation for the molar excess Gibbs energy is given by the sum of

$$\frac{G^E(\text{combinatorial})}{\text{RT}} = \sum_{i=1}^{m} x_i \ln\frac{\Phi_i}{x_i} + \frac{z}{2}\sum_{i=1}^{m} q_i x_i \ln\frac{\theta_i}{\Phi_i} \tag{23}$$

and

$$\frac{G^E(\text{residual})}{\text{RT}} = -\sum_{i=1}^{m} q_i' x_i \ln\left(\sum_{j=1}^{m} \theta_j' \tau_{ji}\right) \tag{24}$$

where segment fraction Φ and area fractions θ and θ' are given by

$$\Phi_i = \frac{x_i r_i}{\sum^{j} x_j r_j}, \qquad \theta_i = \frac{q_i x_i}{\sum_{j=1}^{m} q_j x_j}, \qquad \theta_i' = \frac{q_i' x_i}{\sum_{j=1}^{m} q_j' x_j} \tag{25}$$

The activity coefficient for any component i is given by

$$\ln\gamma_i = \ln\frac{\Phi_i}{x_i} + \frac{z}{2} q_i \ln\frac{\theta_i}{\Phi_i} + l_i - \frac{\Phi_i}{x_i}\sum_{j=1}^{m} x_j l_j - q_i' \ln\left(\sum_{j=1}^{m} \theta_j' \tau_{ji}\right) + q_i' - q_i' \sum_{j=1}^{m} \frac{\theta_j' \tau_{ij}}{\sum_{k=1}^{m} \theta_k' \tau_{kj}} \tag{26}$$

i-j	**u_{ij}(J mol^{-1})**	**u_{ji}(J mol^{-1})**
1–2	3589.99	−2835.91
1–3	909.55	4182.19
2–3	3043.92	−4435.44
*$r_1 = 0.920$, $r_2 = 3.000$, $r_3 = 4.827$		*$q_1 = 1.400$, $q_2 = 4.000$, $q_3 = 4.196$

*r_i and q_i are the UN1QUAC structural parameters.

Table 5. Optimized UNIQUAC binary interaction energy parameters, u_{ij} and u_{ji}(J mol^{-1}), for water (1) + phosphoric acid (2) + butyl acetate (3) ternary system at $T = 308.2$ K [12].

The molecular interaction-based model of UNIQUAC successfully has correlated many LLE data of ternary and quaternary systems [10, 15–17, 24] within Aspen Plus—for example, see **Table 5**; however, as previously mentioned, during the last 10 years, many LLE-based algorithm—GA, SA, and BA—have been developed and introduced in the literature [57, 60, 61]. Recently, a study on *improved binary parameters using GA for ternary mixtures: UNIQUAC model* has been carried out by Hamidi [63].

4.3. GMDH-type neural network

It is observed that usually LLE measurements are successfully correlated using common models such as NRTL and UNIQUAC. However, these conventional thermodynamic methods for data prediction of complex systems are tedious and involve a certain amount of empiricism. Recently, a new prediction method, the group method of data handling, was developed to predict LLE and VLE data in order to avoid these limitations [64, 65]. Namely, the GMDH tie line data for the ternary system of water + phosphoric acid + dichloromethane have been illustrated in **Table 1** [10, 26]. Furthermore, the reliability of the presented tie line results can be tested through the Othmer-Tobias [66] and Bachman [67] correlation equations.

Acknowledgements

We gratefully acknowledge the financial support from the Research Council of Firoozabad Branch, Islamic Azad University.

Author details

Khatereh Bahrpaima

Address all correspondence to: Kh.bahrpyma@iauf.ac.ir

Department of Chemistry, Firoozabad Branch, Islamic Azad University, Firoozabad, Iran

References

[1] Hignett T P. Production of wet process phosphoric acid: Proceedings of the 2nd International Congress on Phosphorous Compounds; Boston, USA; 1980. p. 401–429.

[2] Hamdi R, Hannachi A. Comparative simulation of the purification of wet phosphoric acid by tbp, mibk and a mixture (MIBK+TBP). J. Chem. Eng. Process Technol. 2012;**3**:2. DOI: 10.4172/2157-7048.1000134

[3] Ross W M H, Jones R M, Durgin C B. The purification of phosphoric acid by crystallization. Ind. Eng. Chem. 1925;**17**:1081–1083. DOI: 10.1021/ie50190a031

[4] Kumar B N, Radhika S, Kantam M L, Reddy B R. Solid–liquid extraction of terbium from phosphoric acid solutions using solvent-impregnated resin containing TOPS 99. J. Chem. Technol. Biotechnol. 2011;**86**(4)562–569. DOI: 10.1002/jctb.2553

[5] Koncsag C I, Barbulescu A. Liquid-liquid extraction with and without a chemical reaction. In: El-Amin, Mohamed, (ed.) Mass transfer in multiphase systems and its applications. [Rijeka, Croatia]: InTech; 2011. 207–232 p. ISBN 978-953-307-215-9

[6] Shekarsaraee S. Liquid-liquid equilibrium study for the system (water + phosphoric acid + propylene carbonate) at different temperatures. J. Chem. Thermodyn. 2017;**104**:16–23. DOI: 10.1016/j.jct.2016.09.008

[7] Mohsen-Nia M, Jazi B, Amiri H. Binodal curve measurements for (water + propionic acid + dichloromethane) ternary system by cloud point method. J. Chem. Thermodyn. 2009;**41**:859–863. DOI: 10.1016/j.jct.2009.02.004

[8] Merzougui A, Hasseine A, Kabouche A, Korichi M. LLE for the extraction of alcohol from aqueous solutions with diethyl ether and dichloromethane at 293.15 K, parameter estimation using a hybrid genetic based approach. Fluid Phase Equilib. 2011;**309(2)**:161–167. DOI: 10.1016/j.fluid.2011.07.011

[9] Ghanadzadeh Gilani H, Ghanadzadeh Gilani A, Shekarsaraee S. Experimental study of phase equilibria in aqueous mixtures of phosphoric acid with isoamyl acetate and methyl isoamyl ketone at T = (298.2, 308.2, and 318.2) K. Fluid Phase Equilib. 2013;**337**:32–38. DOI: 10.1016/j.fluid.2012.09.038

[10] Shekarriz S. Prediction of diagram phase of aqueous mixtures of phosphoric acid with dichloromethane and 1,2-dichloroethane using GMDH-type neural network [thesis]. Firoozabad: Firoozabad Branch, Islamic Azad University; 2012.

[11] Doshman ziari S, Bahrpaima Kh, Sharafi Z. (Liquid + liquid) equilibria of quaternary systems containing water, phosphoric acid, dichloromethane and 1,2-dichloroethane. J. Chem. Thermodyn. 2014;**68**:327–331. DOI: 10.1016/j.jct.2013.05.016

[12] Ghanadzadeh H, Ghanadzadeh A, Bahrpaima Kh. Liquid phase equilibria of (water + phosphoric acid + 1-butanol or butyl acetate) ternary systems at T = 308.2 K. J. Chem. Thermodyn. 2008;**40**:1666–1670. DOI: 10.1016/j.jct.2008.07.001

[13] Ghanadzadeh H, Ghanadzadeh A, Aghajani Z, Abbasnejad S, Shekarsaraee S. (Liquid + liquid) equilibria in ternary aqueous mixtures of phosphoric acid with organic solvents at T = 298.2 K. J. Chem. Thermodyn. 2010;**42**:695–699. DOI: 10.1016/j.jct.2010.01.001

[14] Parameswaran A, Paluri B R. Ternary liquid equilibria of the water-phosphoric acid-isoamyl alcohol, cyclohexanol, or methyl isobutyl ketone systems at 35°C. J. Chem. Eng. Data. 1968;**13(2):**194–196. DOI: 10.1021/je60037a015

[15] Gomez-Siurana A, Ruiz-Bevia F, Fernandez-Sempere J, Torregrosa-Fuerte E. Purification of phosphoric acid by extraction with 2-ethyl-1-hexanol: equilibrium data and mass transfer coefficients. Ind. Eng. Chem. Res. 2001;**40**:892–897. DOI: 10.1021/ie000065a

[16] Ghanadzadeh H, Ghanadzadeh A, Shekarsaraee S, Uslu H. Liquid phase equilibria of the system (water + phosphoric acid + 1-octanol) at T = (298.2, 308.2, and 318.2) K. Fluid Phase Equilib. 2012;**316**:109–116. DOI: 10.1016/j.fluid.2011.12.016

[17] Shekarsaraee S. Phase equilibria of the ternary system water + phosphoric acid + 1-nonanol at different temperatures. Phys. Chem. Res. 2016;**4(3)**:507–518.

[18] Ghanadzadeh H, Ghanadzadeh A, Shekarsaraee S. Solubility and tie line data of the water phosphoric acid–solvents at T = 303.2, 313.2, and 323.2 K: An experimental and correlational study. Thermochimica Acta. 2013;**558**:36–45. DOI: 10.1016/j.tca.2013.02.006

[19] Ghanadzadeh H, Ghanadzadeh A, Shekarsaraee S, Nasiri-Touli E, Seyed Saadat S L. Liquid–liquid equilibria study of the (water + phosphoric acid + hexyl or cyclohexyl acetate) systems at T = (298.15, 308.15, and 318.15) K: measurement and thermodynamic modelling. J. Chem. Thermodyn. 2016;**98**:200–207. DOI: 10.1016/j.jct.2016.03.025

[20] Dhouib-Sahnoun R, Feki M, Ayedi H F. Liquid-liquid equilibria of the ternary system water + phosphoric acid + tributyl phosphate at 298.15 K and 323.15 K. J. Chem. Eng. Data. 2002;**47**:861–866. DOI: 10.1021/je010293r

[21] Carlos H H, Pedro R, Liquid-liquid equilibria of the system water/phosphoric acid/ diisopropyl ether at (273.15, 283.15, and 293.15) K. J. Chem. Eng. Data. 2004;**49**:218–220. DOI: 10.1021/je030158+

[22] Marcilla A F, Ruiz F, Sabater M C. Two-phase and three-phase liquid-liquid equilibrium for bis(2-methylpropyl) ether + phosphoric acid + water. J. Chem. Eng. Data. 1994;**39 (1)**:14–18. DOI: 10.1021/je00013a005

[23] Changwen Ye. Liquid–liquid equilibria for the ternary system containing water + phosphoric acid + organic entrainers at 308.15 and 328.15 K. J. Mol. Liquids. 2013;**186**:39–43. DOI: 10.1016/j.molliq.2013.05.006

[24] Ghanadzadeh H, Ghanadzadeh A, Shekarsaraee S, Uslu H. (Liquid + liquid) equilibrium data of (water + phosphoric acid + solvents) systems at T = (308.2 and 318.2) K. J. Chem. Thermodyn. 2012;**53**:52–59. DOI: 10.1016/j.jct.2012.04.011

[25] Ghanadzadeh H, Ghanadzadeh A, Ahmadifar S. Experimental and correlational study of phase equilibria in aqueous mixtures of phosphoric acid with aromatic hydrocarbons at various temperatures. J. Chem. Thermodyn. 2015;**91**:121–126. DOI: 10.1016/j.jct.2015.07.020

[26] Bahrpaima Kh, Ebrahimi M, Madani H, Shekarriz S. Salting-out effect on the (liquid + liquid) equilibrium for the ternary systems (water + phosphoric acid + dichloromethane or 1,2-dichloroethane). J. Chem. Thermodyn. 2014;**71**: 118–125. DOI: 10.1016/j.jct.2013.11.034

[27] Muralimohan S, Bhimeswara R P. Ternary liquid equilibria of the water-phosphoric acid-1-butanol, butyl acetate, or methyl ethyl ketone systems at 35 °C. J. Chem. Eng. Data. 1967;**12(4)**:494–497. DOI: 10.1021/je60035a008

[28] Santos F S, d'Avila S G, Aznar M. Salt effect on liquid–liquid equilibrium of water + 1-butanol + acetone system: experimental determination and thermodynamic modeling. Fluid Phase Equilib. 2001;**187–188**:265–274. DOI: 10.1016/S0378-3812(01)00541-6

[29] Aznar M, Araujo R N, Romanato J F, Santos G R, d'Avila S G. Salt effects on liquid−liquid equilibrium in water + ethanol + alcohol + salt systems. J. Chem. Eng. Data. 2000;**45 (6)**:1055–1059. DOI: 10.1021/je000029i

[30] Wannachod T, Hronec M, Sotak T, Fulajtarova K, Pancharoen U, Nootong K. Influence of salt on the solubility and tie-line data for water + formic acid + methyl isobutyl ketone at T = 298.15 K. Chem. Eng. Data. 2016;**61(7)**:2433–2439. DOI: 10.1021/acs.jced.6b00109

[31] Ghalami-Choobar B, Ghanadzadeh A, Kousarimehr S. Salt effect on the liquid-liquid equilibrium of (water + propionic acid + cyclohexanol) system at T= (298.2, 303.2, and 308.2) K. Chin. J. Chem. Eng. 2011;**19(4)**:565–569. DOI: 10.1016/S1004-9541(11)60022-0

[32] Govindarajan M, Sabarathinam P. Effect of some inorganic salts on the ternary liquid−liquid equilibria of the water + 4-methyl-2-pentanone + propanoic or butanoic acid at 35 °C. J. Chem. Eng. Data. 1997;**42(2)**:402–408 DOI: 10.1021/je960281j

[33] Vakili-Nezhaad G R, Mohsen-Nia M, Taghikhani V, Behpoor M, Aghahosseini M. Salting-out effect of NaCl and KCl on the ternary LLE data for the systems of (water + propionic acid + isopropyl methyl ketone) and of (water + propionic acid + isobutyl methyl ketone). J. Chem. Thermodyn. 2004;**36**:341–348. DOI: 10.1016/j.jct.2003.11.011

[34] Santos F S, d'Avila S G, Aznar M. Salt effect of KBr on the liquid-liquid equilibrium of the water/ethanol/1-pentanol system. Braz. J. Chem. Eng. 2000;**17(4-7)**:721–734. DOI: 10.1590/S0104-66322000000400036

[35] Ruiz F, Galan M I, Boluda N. Quaternary liquid–liquid equilibrium: water–phosphoric acid–1-butanol–2-butanone at 25°C. Fluid Phase Equilib. 1998;**146(1–2)**;175–185.

[36] Ruiz F, Prats D, Gomis V. Quaternary liquid-liquid equilibrium: water-ethanol-chloroform-toluene at 25°C. Experimental determination and graphical and analytical correlation of equilibrium data. J. Chem. Eng. Data. 1985;**30(4)**:412–416. DOI: 10.1021/je00042a013

[37] Chen Y, Fu M, Cao C, Chen E. Quaternary liquid–liquid equilibria for aqueous systems containing dimethyl carbonate at 298.15 K. J. Solut. Chem. 2008;**37(11)**:1529–1540. DOI: 10.1007/s10953-008-9325-8

[38] Ruiz F, Gomis V. Correlation of quaternary liquid-liquid equilibrium data using UNIQUAC. Ind. Eng. Chem. Process Des. Dev. 1986;**25(1)**;216–220. DOI: 10.1021/i200032a035

[39] Mohsen-Nia M, Paikar I. (Liquid + liquid) equilibria of ternary and quaternary systems containing n-hexane, toluene, m-xylene, propanol, sulfolane, and water at T = 303.15 K. J. Chem. Thermodyn. 2007;**39**:1085–1089. DOI:10.1016/j.jct.2006.12.008

[40] Mohammad Doulabi F S, Mohsen-Nia M, Modarress H. Measurements and modeling of quaternary (liquid + liquid) equilibria for mixtures of (methanol or ethanol + water + toluene + n-dodecane). J. Chem. Thermodyn. 2006;**38**:405–412. DOI:10.1016/j.jct.2005.06.007

[41] Mohsen-Nia M. Experimental and theoretical study of quaternary (liquid + liquid) equilibria for mixtures of (methanol or water + ethanol + toluene + n-decane). J. Chem. Thermodyn. 2006;**38**:1285–1291. DOI:10.1016/j.jct.2006.02.011

[42] Liu C, Ren Y, Wang Y. Liquid−liquid equilibria for the quaternary system H_3PO_4−NaCl−H_2O−TBP at 298.15 K. J. Chem. Eng. Data. 2014;**59(1)**:70–75. DOI: 10.1021/je400817m

[43] Bahrpaima Kh, Bijanzadeha A R, Behzadi M. Effect of magnetic field on the liquid-liquid equilibria of (water + acetic acid + organic solvent) ternary systems. Phys. Chem. Res. 2017;**5(1)**:125–134. DOI: 10.22036/pcr.2017.38854

[44] Yongli S, Yong L, Songhai W, Shaoyi J. Effect of magnetic field on the extraction process of acetone-water-trichloroethane system, Chin. J. Chem. Eng. 2007;**15(6)**:916–918.

[45] Mohsen-Nia M, Jazi B, Amiri H. Effects of external electromagnetic field on binodal curve of (water + propionic acid + dichloromethane) ternary system. J. Chem. Thermodyn. 2009;**41**:1081–1085. DOI: 10.1016/j.jct.2009.04.013

[46] Shouli G. Liquid–liquid equilibrium of (water (1) + acetic acid (2) + cyclohexane (3)) ternary system under ultrasonic wave [thesis]. Firoozabad: Firoozabad Branch, Islamic Azad University; 2011.

[47] Hossieni A. Effect of ultrasonic wave on equilibria phase for the ternary system of water (1) + acetic acid (2) + cyclohexanol (3) [thesis]. Firoozabad: Firoozabad Branch, Islamic Azad University; 2011.

[48] Attarzadeh M. Study of diagram phase for the ternary system of (water + acetic acid + cyclohexanone) under ultrasonic wave [thesis]. Firoozabad: Firoozabad Branch, Islamic Azad University; 2011.

[49] Reichardt C. Solvents and Solvent Effects in Organic Chemistry, 3rd ed., Wiley-VCH; Germany, 2003.

[50] Kamlet M J, Abboud J M, Abraham M H, Taft R W. Linear solvation energy relationships. 23. A comprehensive collection of the solvatochromic parameters, π^*, α, and β, and some methods for simplifying the generalized solvatochromic equation. J. Org. Chem. 1983;**48**; 2877–2887. DOI: 10.1021/jo00165a018

[51] Katritzky A R, Fara D C, Yang H, Taemm K, Tamm T, Karelson M. Quantitative measures of solvent polarity. Chem. Rev. 2004;**104(1)**:175–198. DOI: 10.1021/cr020750m

[52] Renon H, Prausnitz J M. Local compositions in thermodynamic excess functions for liquid mixtures. AIChE J. 1968;**14**:135–144, DOI: 10.1002/aic.690140124

[53] Abrams D S, Prausnitz J M. Statistical thermodynamics of liquid mixtures: A new expression for the excess Gibbs energy of partly or completely miscible systems. AIChE J. 1975;**21(1)**:116–128. DOI: 10.1002/aic.690210115

[54] Wilson G M, Deal C H. Activity coefficient and molecular structure. Ind. Eng. Chem. Fundam. 1962;**1(1)**:20–23. DOI: 10.1021/i160001a003

[55] Anderson G M, Crerar D A. Thermodynamics in Geochemistry: The Equilibrium Model. USA: Oxford University Press; 1993.

[56] Prausnitz J M, Lichtenthaler R N, de Azevedo E G. Molecular Thermodynamics of Fluid-Phase Equilibria, 3rd ed. Prentice Hall; The United States of America, 1998.

[57] Singh M K, Banerjee T, Khanna A. Genetic algorithm to estimate interaction parameters of multicomponent systems for liquid–liquid equilibria. Comput. Chem. Eng. 2005;**29**:1712–1719. DOI: 10.1016/j.compchemeng.2005.02.020

[58] Pham D T, Ghanbarzadeh A, Koc E, Otri S, Rahim S, Zaidi M. The bees algorithm—a novel tool for complex optimisation problems. In: Proceedings of the 2nd International Virtual Conference on Intelligent Production Machines and Systems (IPROMS 2006); 2006. pp. 454–459.

[59] Bertsimas D, Tsitsiklis J. Simulated annealing. Statist. Sci. 1993;**8(1)**:10–15. DOI:10.1214/ss/1177011077

[60] Janamiri L. Calculation of the interaction parameters using simulated annealing algorithm for multi-component mixtures: NRTL model [thesis]. Firoozabad: Firoozabad Branch, Islamic Azad University; 2014.

[61] Rezaei F. Calculation of the interaction parameters using simulated bee algorithm for multicomponent mixtures: NRTL model [thesis]. Firoozabad: Firoozabad Branch, Islamic Azad University; 2015.

[62] Anderson N. On the Calculation of Filter Coefficients for Maximum Entropy Spectral Analysis. In Childers: Modern Spectrum Analysis, IEEE Press; New York, 1978.

[63] Hamidi M. Improved binary parameters using GA for ternary mixtures: UNIQUAC model [thesis]. Firoozabad: Firoozabad Branch, Islamic Azad University; 2014.

[64] Reyhani S Z, Ghanadzadeh H, Puigjaner L, Recances F. Estimation of liquid−liquid equilibrium for a quaternary system using the GMDH algorithm. Ind. Eng. Chem. Res. 2009;**48(4)**:2129–2134. DOI: 10.1021/ie801082s

[65] Ketabchi S, Ghanadzadeh H, Ghanadzadeh A, Fallahi S, Ganji M. Estimation of VLE of binary systems (*tert*-butanol + 2-ethyl-1-hexanol) and (n-butanol + 2-ethyl-1-hexanol) using GMDH-type neural network. J. Chem. Thermodyn. 2010;**42(11)**:1352–1355. DOI: 10.1016/j.jct.2010.05.018

[66] Othmer D F, Tobias P E. Liquid-liquid extraction data-tie-line correlation. Ind. Eng. Chem. 1942;**34(6)**:693–696, DOI: 10.1021/ie50390a600.

[67] Bachman I. Tie lines in ternary liquid systems. Ind. Eng. Chem. Anal. Ed. 4012;**19(1)**:38–39. DOI: 10.1021/ac50141a013